My 132 Semesters of Chemistry Studies

My 132 Semesters of Chemistry Studies

Studium chymiae nec nisi cum morte finitur

Vladimir Prelog

Translated from the German by Otto Theodor Benfey and David Ginsburg

PROFILES, PATHWAYS, AND DREAMS
Autobiographies of Eminent Chemists

Jeffrey I. Seeman, Series Editor

American Chemical Society, Washington, DC 1991

Library of Congress Cataloging-in-Publication Data

Prelog, V.
  My 132 Semesters of Chemistry Studies: Studium chymiae
nec nisi cum morte finitur
    p.   cm.—(Profiles, pathways, and dreams, ISSN 1047–8329)
  Includes bibliographical references and index.
  ISBN 0–8412–1772–6 (cloth)
  ISBN 0–8412–1798–X (pbk.)

  1. Prelog, V.   2. Chemists—Yugoslavia—Biography.
3. Chemistry, Organic—History—20th century. I. Title.
II. Title: My one hundred thirty two semsters of
chemistry studies. III. Series.

QD22.P73A3   1991
540'.92—dc20   90–951

The paper used in this publication meets the minimum requirements of American National Standard for Information Sciences—Permanence of Paper for Printed Library Materials, ANSI Z39.48–1984.

∞

Copyright © 1991

American Chemical Society

All Rights Reserved. The copyright owner consents that reprographic copies may be made for personal or internal use or for the personal or internal use of specific clients. This consent is given on the condition, however, that the copier pay the stated per-copy fee through the Copyright Clearance Center, Inc., 27 Congress Street, Salem, MA 01970, for copying beyond that permitted by Sections 107 or 108 of the U.S. Copyright Law. This consent does not extend to copying or transmission by any means—graphic or electronic—for any other purpose, such as for general distribution, for advertising or promotional purposes, for creating a new collective work, for resale, or for information storage and retrieval systems. The copying fee is $0.75 per page. Please report your copying to the Copyright Clearance Center with this code: 1047–8329/91/$00.00+.75.

The citation of trade names and/or names of manufacturers in this publication is not to be construed as an endorsement or as approval by ACS of the commercial products or services referenced herein; nor should the mere reference herein to any drawing, specification, chemical process, or other data be regarded as a license or as a conveyance of any right or permission to the holder, reader, or any other person or corporation, to manufacture, reproduce, use, or sell any patented invention or copyrighted work that may in any way be related thereto. Registered names, trademarks, etc., used in this publication, even without specific indication thereof, are not to be considered unprotected by law.

PRINTED IN THE UNITED STATES OF AMERICA

1991 ACS Books Advisory Board

V. Dean Adams
Tennessee Technological
　University

Paul S. Anderson
Merck Sharp & Dohme
　Research Laboratories

Alexis T. Bell
University of California—Berkeley

Malcolm H. Chisholm
Indiana University

Natalie Foster
Lehigh University

Dennis W. Hess
University of California—Berkeley

Mary A. Kaiser
E. I. du Pont de Nemours and
　Company

Gretchen S. Kohl
Dow-Corning Corporation

Michael R. Ladisch
Purdue University

Bonnie Lawlor
Institute for Scientific Information

John L. Massingill
Dow Chemical Company

Robert McGorrin
Kraft General Foods

Julius J. Menn
Plant Sciences Institute,
　U.S. Department of Agriculture

Marshall Phillips
Office of Agricultural Biotechnology,
　U.S. Department of Agriculture

Daniel M. Quinn
University of Iowa

A. Truman Schwartz
Macalaster College

Stephen A. Szabo
Conoco Inc.

Robert A. Weiss
University of Connecticut

Foreword

In 1986, the ACS Books Department accepted for publication a collection of autobiographies of organic chemists, to be published in a single volume. However, the authors were much more prolific than the project's editor, Jeffrey I. Seeman, had anticipated, and under his guidance and encouragement, the project took on a life of its own. The original volume evolved into 22 volumes, and the first volume of Profiles, Pathways, and Dreams: Autobiographies of Eminent Chemists was published in 1990. Unlike the original volume, the series was structured to include chemical scientists in all specialties, not just organic chemistry. Our hope is that those who know the authors will be confirmed in their admiration for them, and that those who do not know them will find these eminent scientists a source of inspiration and encouragement, not only in any scientific endeavors, but also in life.

M. Joan Comstock
Head, Books Department
American Chemical Society

Contributors

We thank the following corporations and Herchel Smith for their generous financial support of the series Profiles, Pathways, and Dreams.

Akzo nv

Bachem Inc.

E. I. du Pont de Nemours and Company

Duphar B.V.

Eisai Co., Ltd.

Fujisawa Pharmaceutical Co., Ltd.

Hoechst Celanese Corporation

Imperial Chemical Industries PLC

Kao Corporation

Mitsui Petrochemical Industries, Ltd.

The NutraSweet Company

Organon International B.V.

Pergamon Press PLC

Pfizer Inc.

Philip Morris

Quest International

Sandoz Pharmaceuticals Corporation

Sankyo Company, Ltd.

Schering–Plough Corporation

Shionogi Research Laboratories, Shionogi & Co., Ltd.

Herchel Smith

Suntory Institute for Bioorganic Research

Takasago International Corporation

Takeda Chemical Industries, Ltd.

Unilever Research U.S., Inc.

Profiles, Pathways, and Dreams

Titles in This Series

Sir Derek H. R. Barton *Some Recollections of Gap Jumping*

Arthur J. Birch *To See the Obvious*

Melvin Calvin *Following the Trail of Light: A Scientific Odyssey*

Donald J. Cram *From Design to Discovery*

Michael J. S. Dewar *A Semiempirical Life*

Carl Djerassi *Steroids Made It Possible*

Ernest L. Eliel *From Cologne to Chapel Hill*

Egbert Havinga *Enjoying Organic Chemistry, 1927–1987*

Rolf Huisgen *Mechanisms, Novel Reactions, Synthetic Principles*

William S. Johnson *A Fifty-Year Love Affair with Organic Chemistry*

Raymond U. Lemieux *Explorations with Sugars: How Sweet It Was*

Herman Mark *From Small Organic Molecules to Large: A Century of Progress*

Bruce Merrifield *The Concept and Development of Solid-Phase Peptide Synthesis*

Teruaki Mukaiyama *To Catch the Interesting While Running*

Koji Nakanishi *A Wandering Natural Products Chemist*

Tetsuo Nozoe *Seventy Years in Organic Chemistry*

Vladimir Prelog *My 132 Semesters of Chemistry Studies*

John D. Roberts *The Right Place at the Right Time*

Paul von Rague Schleyer *From the Ivy League into the Honey Pot*

F. G. A. Stone *Organometallic Chemistry*

Andrew Streitwieser, Jr. *A Lifetime of Synergy with Theory and Experiment*

Cheves Walling *Fifty Years of Free Radicals*

About the Editor

JEFFREY I. SEEMAN received his B.S. with high honors in 1967 from the Stevens Institute of Technology in Hoboken, New Jersey, and his Ph.D. in organic chemistry in 1971 from the University of California, Berkeley. Following a two-year staff fellowship at the Laboratory of Chemical Physics of the National Institutes of Health in Bethesda, Maryland, he joined the Philip Morris Research Center in Richmond, Virginia, where he is currently a senior scientist and project leader. In 1983–1984, he enjoyed a sabbatical year at the Dyson Perrins Laboratory in Oxford, England, and claims to have visited more than 90% of the castles in England, Wales, and Scotland.

Seeman's 80 published papers include research in the areas of photochemistry, nicotine and tobacco alkaloid chemistry and synthesis, conformational analysis, pyrolysis chemistry, organotransition metal chemistry, the use of cyclodextrins for chiral recognition, and structure–activity relationships in olfaction. He was a plenary lecturer at the Eighth IUPAC Conference on Physical Organic Chemistry held in Tokyo in 1986 and has been an invited lecturer at numerous scientific meetings and universities. Currently, Seeman serves on the Petroleum Research Fund Advisory Board. He continues to count Nero Wolfe and Archie Goodwin among his best friends.

Contents

List of Photographs	xiii
Preface	xv
Editor's Note	xix
My 132 Semesters of Chemistry Studies	1
Childhood and Youth in Yugoslavia (1906–1924)	3
Student of Chemistry in Prague (1924–1929)	7
Practicing Chemistry in Prague (1929–1934)	12
University of Zagreb (1935–1941)	13
Zurich Since 1942	20
Lectureship at Notre Dame (1950)	45
Full Professor in Zurich, Lectureship at Columbia University (1951): Revolution in Instrumentation and Other Developments (1950–1957)	47

Laboratory Head (1957), Collegial Chairmanship (1964),
and Retirement (1976) 68

In Retrospect 85

Epilogue 87

References 97

Index 107

Photographs

The Buergenstock Declaration, 1966	xx
With Albert Eschenmoser at Bürgenstock, 1989	xxii
The famous CIP trio	xxiii
The author with his parents, in Sarajevo, 1908	4
In Sarajevo, 1912	5
Experiment designed for the photographer, 1918	6
Rudolph Lukeš (1897–1960)	9
A sublieutenant in the Yugoslav Royal Navy, 1934	13
First-day cover commemorating the Czechoslovakian Chemical Society Centennial	21
With Leopold Ruzicka	27
Sir Robert and Lady Robinson in Paris, 1957	29
Maurice Marie Janot (1903–1978)	32
With R. B. Woodward at the CIBA Foundation, London, 1977	35
Sir Derek H. R. Barton	36
With Karel Wiesner in Fredericton, New Brunswick, 1972	44

With Charles C. Price at Notre Dame, 1950	45
Placidus Andreas Plattner (1904–1975) in Zurich	48
With Kamila Prelog, Cambridge, Massachusetts, 1951	49
Rita and Sir John ("Kappa") Cornforth	65
With Duilio Arigoni and Frank Westheimer, Bürgenstock, 1976	69
Oskar Jeger, Jack Dunitz, and Albert Eschenmoser, 1986	70
Admiring the A. W. Hofmann Medal with Kamila Prelog, 1967	71
With chemistry students in Lindau, 1977	72
Magic with Koji Nakanishi at IUPAC, 1984	73
With Ernest L. Eliel at the Nomenclature Conference, 1968	74
Robert Sidney Cahn at Bürgenstock, 1966	76
Sir Christopher Ingold at Bürgenstock, 1966	77
Yuri Ovchinnikov at Bürgenstock, 1966	81
Wilhelm Simon in Zurich	84
With Leopold Ruzicka in Zurich, 1975	85
At Bürgenstock, 1989	86
Ruzicka's ex libris by Hans Erni	87
The author's ex libris by Hans Erni	88
With Robert Woodward and Carl Djerassi in the Baltic Sea, 1970	88–89

Preface

"How did you get the idea—and the good fortune—to convince 22 world-famous chemists to write their autobiographies?" This question has been asked of me, in these or similar words, frequently over the past several years. I hope to explain in this preface how the project came about, how the contributors were chosen, what the editorial ground rules were, what was the editorial context in which these scientists wrote their stories, and the answers to related issues. Furthermore, several authors specifically requested that the project's boundary conditions be known.

As I was preparing an article[1] for *Chemical Reviews* on the Curtin–Hammett principle, I became interested in the people who did the work and the human side of the scientific developments. I am a chemist, and I also have a deep appreciation of history, especially in the sense of individual accomplishments. Readers' responses to the historical section of that review encouraged me to take an active interest in the history of chemistry. The concept for Profiles, Pathways, and Dreams resulted from that interest.

My goal for Profiles was to document the development of modern organic chemistry by having individual chemists discuss their roles in this development. Authors were not chosen to represent my choice of the world's "best" organic chemists, as one might choose the "baseball all-star team of the century". Such an attempt would be foolish: Even the selection committees for the Nobel prizes do not make their decisions on such a premise.

The selection criteria were numerous. Each individual had to have made seminal contributions to organic chemistry over a multidecade career. (The average age of the authors is over 70!) Profiles would represent scientists born and professionally productive in different countries. (Chemistry in 13 countries is detailed.) Taken together, these individuals were to have conducted research in nearly all subspecialties of organic chemistry. Invitations to contribute were based on solicited advice and on recommendations of chemists from five continents, including nearly all of the contributors. The final assemblage was selected entirely and exclusively by me. Not all who were invited chose to participate, and not all who should have been invited could be asked.

A very detailed four-page document was sent to the contributors, in which they were informed that the objectives of the series were

1. to delineate the overall scientific development of organic chemistry during the past 30–40 years, a period during which this field has dramatically changed and matured;

2. to describe the development of specific areas of organic chemistry; to highlight the crucial discoveries and to examine the impact they have had on the continuing development in the field;

3. to focus attention on the research of some of the seminal contributors to organic chemistry; to indicate how their research programs progressed over a 20–40-year period; and

4. to provide a documented source for individuals interested in the hows and whys of the development of modern organic chemistry.

One noted scientist explained his refusal to contribute a volume by saying, in part, that "it is extraordinarily difficult to write in good taste about oneself. Only if one can manage a humorous and light touch does it come off well. Naturally, I would like to place my work in what I consider its true scientific perspective, but..."

Each autobiography reflects the author's science, his lifestyle, and the style of his research. Naturally, the volumes are not uniform, although each author attempted to follow the guidelines. "To write in good taste" was not an objective of the series. On the contrary, the authors were specifically requested not to write a review article of their field, but to detail their own research accomplishments. To the extent that this instruction was followed and the result is not "in good taste", then these are criticisms that I, as editor, must bear, not the writer.

As in any project, I have a few regrets. It is truly sad that Egbert Havinga, who wrote one volume, and David Ginsburg, who translated another, died during the development of this project. There have been many rewards, some of which are documented in my personal account of this project, entitled "Extracting the Essence: Adventures of an Editor" published in *CHEMTECH*.[2]

Acknowledgments

I join the entire chemical community in offering each author unbounded thanks. I thank their families and their secretaries for their contributions. Furthermore, I thank numerous chemists for reading and reviewing the autobiographies, for lending photographs, for sharing information, and for providing each of the authors and me the encouragement to proceed in a project that was far more costly in time and energy than any of us had anticipated.

I thank my employer, Philip Morris USA, and J. Charles, R. N. Ferguson, K. Houghton, and W. F. Kuhn, for without their support Profiles, Pathways, and Dreams could not have been. I thank ACS Books, and in particular, Robin Giroux (acquisitions editor), Karen Schools Colson (production manager), Janet Dodd (senior editor), Joan Comstock (department head), and their staff for their hard work, dedication, and support. Each reader no doubt joins me in thanking 24 corporations and Herchel Smith for financial support for the project.

I thank my children, Jonathan and Brooke, for their patience and understanding; remarkably, I have been working on Profiles for more than half of their lives—probably the only half that they can remember! Finally, I again thank all those mentioned and especially my family, friends, colleagues, and the 22 authors for allowing me to share this experience with them.

JEFFREY I. SEEMAN
Philip Morris Research Center
Richmond, VA 23234

November 11, 1990

[1] Seeman, J. I. *Chem. Rev.* **1983**, *83*, 83–134.
[2] Seeman, J. I. *CHEMTECH* **1990**, *20*(2), 86–90.

Editor's Note

Vladimir Prelog laughed softly and contagiously as he told of the student who approached him one day at a university in the United States. "I know you," the young man exclaimed proudly, "You're Djerassi–Prelog!" Prelog thoroughly enjoyed the humor to be found in the fact that he was so exuberantly acknowledged for one of his minor contributions, the *Prelog–Djerassi lactone* and its related *Prelog–Djerassi lactonic acid*.

Now in his ninth decade, Prelog has achieved so much: a Nobel Prize; a professorship at a major institution; newspaper and magazine articles about him; the naming in his honor of a seminal nomenclature system (the Cahn–Ingold–Prelog, or CIP, system); and the aforementioned oddly substituted lactone. Prelog, by any measure, has reached the pinnacle of scientific accomplishment and recognition.

His charm and wit are legendary, as demonstrated by his mirror-image signature on the declaration on pages xx and xxi; yet he answers his phone with an unassuming "Prelog". Because of university regulations requiring formal affiliation with the school in order to maintain an office, Prelog remains active daily as a "postdoctoral student" at the ETH in Zurich. The nonagenarian, with his gentle, self-deprecating sense of humor, has chosen this role to identify himself in the title for

Buergenstock Declaration

<u>Whereas</u> we the stereochemists of many lands, being gathered together in solemn conclave at the Buergenstock in the Half-Canton of Nidwalden belonging to the Helvetian Confederation, do freely recognize, agree, affirm, and proclaim that the RULES of CAHN, INGOLD and PRELOG, as set down in their articles, are just rules and fair, causing harm to no man,

<u>Each</u> of us, being of sound mind, does freely agree

1. to observe the aforesaid RULES in all such cases as they may be applicable, and

2. to refrain from using other rules as may have been or as may be proposed by other powers.

<u>Whereupon</u> we do affirm that any infringement of this Agreement shall be punished, in the first instance by a penalty amounting to 1 (one) glass of liquid refreshment to be supplied at the costs of the infringer to all stereochemists present; and in further instances by the infringer being compelled to <u>read</u> the aforesaid

articles by CAHN, INGOLD and PRELOG in their entirety.

Signed, this day, Friday, 13th May, 1966, by

A. Dreiding, model-maker, resident in Erlenbach, Switzerland

K. Mislow, chirosophist, resident in Princeton, New Jersey

and witnessed by

R. S. Cahn   Sir Christopher Ingold   V. Prelog
R.S. Cahn   *Ch Ingold*   *prelog V*

Kurt Mislow   as a model
chirosophist   to critically minded
                             dissidents
                                    André Dreiding
                                    stereosophist
                                    stereotactician

The Buergenstock Declaration was written and signed by K. Mislow and A. Dreiding on Friday the 13th of May, 1966, at the Second Bürgenstock Conference. Cahn, Ingold, and Prelog served as witnesses. (Note Prelog's mirror-image signature.) On that day, Ingold had served as moderator for three speakers—H. Eggerer, J. A. Berson, and R. B. Woodward. The title of Woodward's talk was The Maintenance of Orbital Symmetry. (Reprinted with permission from Euchem Conference on Stereochemistry, Bürgenstock, Switzerland, 1965–1989. Copyright 1989 Salle and Sauerländer.)

this volume. Modestly, he accepted the challenge of writing his memoirs because "your list of contributors (to this series)—most of them are my personal friends, so I cannot afford not to join them." This modesty and a distinct sense of shyness are revealed in his writing: "I must keep a little textile . . . Man must be free to hide himself carefully . . . I am practical . . . I have tried to find good (worthy causes), tried to use my efforts to help the community, my family, my university, my city, and my country."

Every day, Prelog generated a few pages in longhand. "Imperfect" they were to him, despite careful readings by his close friends and colleagues, Albert Eschenmoser, Jack Dunitz, and Ernest Eliel. Many times particular sentences or phrases were deleted from one draft to the next as "inappropriate" after being given the most careful scrutiny. Nonetheless, the story "came from my heart. It requires a touch of the poet, one who knows his own weaknesses." His conservatism, precision, and scrutiny also extended to the photographs he selected. (Interestingly, some of those he rejected for his own book he graciously permitted his colleagues to use for their volumes in the *Profiles* series.) When it came time to choose which would be in this volume, he declined to include some of the most captivating; e.g.,

With Albert Eschenmoser at the 25th Bürgenstock Conference in June 1989. (Photograph courtesy J. Seeman.)

The famous CIP trio: R. S. Cahn, whose initials correspond to the (R,S) nomenclature of the CIP system, Sir Christopher Ingold, and Vladimir Prelog at the 1966 Bürgenstock Conference. (Photograph courtesy A. Horeau.)

a photo of him as a 12-year-old chemist was rejected because it was "posed for the photographer" and not "authentic". I argued for its inclusion and Prelog offered me a choice: keep the original (now more than 70 years old) if it were not published, or use it and return it to him. My decision was immediate; the photo appears on page 6. When I reminded him recently of this discussion, he presented me with the treasured photo, saying, "For many years I have collected things; now I am distributing them."

In preparing his manuscript Prelog focused on his chemistry and, despite urging, would relate only a few of the innumerable anecdotes for which he is well known. His concern was that the stories would not be a perfect representation of the truth, whereas the chemistry could be precisely described. It is for this reason that the stories he has chosen to share are found at the end of this text (page 87).

Prelog's modest view of his linguistic skills posed a threat to his participation in this project. Even though his English is superb in terms of usage, vocabulary, and breadth of expression, he was unwilling to draft his manuscript in English. It was resolved that he would prepare his text in German and call upon his friend David Ginsburg to translate it.

The parallel between Prelog's life and that of his mentor and godfather, Leopold Ruzicka, can scarcely go unnoticed: Both were Croatian by birth, both moved to Switzerland shortly before or during a world war, each became a professor of organic chemistry at the ETH, each was involved in natural products research, and each received the Nobel Prize in chemistry. Prelog chose to take refuge at the ETH and collaborate with Ruzicka, in part due to their shared Croatian heritage. It is worth noting that Prelog has spent much time organizing the Ruzicka papers.

Recently, Prelog mused upon his life and autobiography. "Two hundred fifty names of my friends and colleagues are in the index (references). They each now appear before my eyes . . . "

April 16, 1991

Special Acknowledgments to . . .

David Ginsburg, the Technion, Haifa, Israel. Dr. Ginsburg, a noted organic chemist and good friend of Professor Prelog, volunteered to translate the manuscript. Before he began his task, Ginsburg suffered a serious heart attack, yet he insisted upon proceeding as he recuperated. At that time he wrote me that "I am glad that I did it, despite the rate of work; it saved me at a time when I was housebound completely." Dr. Ginsburg died before completing the translation.

Otto Theodor Benfey, National Foundation for the History of Chemistry, Philadelphia, PA. Editor, historian and translator, fluent in German, he, like Prelog, is a physical organic chemist. Dr. Benfey took his doctorate with Christopher Ingold and has had postdoctoral experience with Louis Hammett and Frank Westheimer. Benfey accepted and met the challenge of completing the translation within a very tight deadline.

Jim Tidwell, who prepared the silhouette of the author.

To each of them, our gratitude.

My 132 Semesters of Chemistry Studies

Childhood and Youth in Yugoslavia (1906–1924)

I was born on July 23, 1906, in Sarajevo, which at the time was the capital of the Austro–Hungarian province of Bosnia–Herzegovina. The monarchy was a medley of very different landscapes, peoples, languages, religions, and cultures. This blend left its strong imprint, particularly on the intelligentsia. My paternal grandmother was an Austrian from Schwechat near Vienna; my maternal grandfather was a descendant of 19th-century immigrant builders and masons from San Giovanni di Manzano in Northern Italy. Other ancestors were Croatian farmers and craftsmen. Bosnia and Herzegovina were captured by Austria–Hungary in 1878 during the war against Turkey and were annexed 30 years later. In Sarajevo, people of a variety of different religions lived in almost totally separate communities largely because of religious and societal barriers: the Roman Catholic Croats, the Greek Orthodox Serbs, the indigenous Moslems (who had converted to Islam during Turkish rule), and, last but not least, the Spanioles, Spanish Orthodox Jews who had fled the inquisition during the 15th century.

    The Austro–Hungarian central government took pains that this province, which was neglected under Ottoman rule, would develop materially and culturally and attain a higher level of civilization. For this purpose, the government tried to lure young and qualified Croats from Croatia, Slavonia, and Dalmatia to settle in the new province by various inducements. Thus in 1905, my father came as a young high-school teacher from Croatia to Sarajevo, where I was born.

    I have very vivid memories of two events from the years of my youth in Sarajevo: the murder of the crown prince, Archduke Franz Ferdinand, and his wife on June 28, 1914, and the declaration of war against Serbia a few weeks later. On the occasion of the crown prince's visit, I stood, as a schoolboy, in a cordon in front of the crowds waiting for their Highnesses to drive by. My task was to scatter the contents of a small basket of flowers in front of their carriage. A few hundred meters ahead, the fatal shots were fired. A particularly strong impression was left on me by the subsequent demonstrations permitted by the authorities, during which the mob plundered and set fire to the shops of the Serbian merchants, because

With mother, Mara, née Cettolo (1887–1979), and father, Milan (1879–1931), in Sarajevo in 1908.

all Serbs were considered responsible for the assassination. Since then, I have been allergic to all violent mass demonstrations, even when held for just causes.

Despite the tense political situation, we and another family who were friends of ours went to a summer resort where, on the eve of August 1, 1914, two gendarmes brought the news of the declaration of war on Serbia and the general mobilization. For Austria–Hungary, the war on the Serbian front began badly. The Serbian army moved to the vicinity of Sarajevo, and the thunder of cannons could be heard.

In Sarajevo in 1912.

When my parents separated in 1915, I moved to the Croatian capital of Zagreb, in the care of my father's unmarried sister and his mother, where I stayed for four years. In Zagreb, I attended the first three years of high school (*Realgymnasium*). My aunt was a teacher who was trained in the spirit of the Enlightenment of the 19th century, whose ideals were truth, beauty, and goodness. It was hoped that these ideals could be actualized through science and the arts. My aunt brought me up in the same spirit, and I owe her very much. As a 12-year-old youth in my new home, I was allowed to conduct my first chemical experiments: I distilled and crystallized, prepared oxygen and hydrogen, and, from time to time, caused some damage to the furnishings. There were no chemical kits then for children in Zagreb, but German directions for chemical experiments "for young people" were available. I bought simple implements,

Experiment designed for the photographer in Zagreb in 1918.

such as Bunsen burners, glassware, and filter paper, from a store that had them for chemistry students at the university. I procured the chemicals from the bigger pharmacies. Because I lived in a well-ordered feminine household and because my father was exempted from military service, I did not find the years of war and its privations painful.

In the fall of 1918, after the defeat, Austria–Hungary fell apart, and Croatia became part of the kingdom of Yugoslavia. The unification of the southern Slavic peoples into one country, at first greeted by the population as liberation, did not proceed painlessly. The cultural, religious, and other historically determined differences, as well as the economic and social problems of the various regions, stood in the way of the monarchist government's pressure to force the formation of a unified, centrally ruled state. This situation soon led to dissatisfaction and to profound antagonisms, which were further aggravated by international developments between the two world wars.

In 1918, my father was appointed head of a girls' high school in Osijek, and I soon followed him there. The two years I attended the science-based high school in Osijek deserve mention, because I found in that school a chemistry teacher, Ivan Kuria, who strongly encouraged my love of chemistry. I helped him outfit a teaching laboratory with very simple means, and in the process, I learned much from him (such as glass blowing) that I could make good use of later. Under his supervision, as a 15-year-old student, I wrote my first paper, which appeared in the respected *Chemiker Zeitung*.[1] Its publication is an indication of the low standard of the chemical literature at that time. In 1921, my father moved to Zagreb, where he became a professor of modern history at the university. In a country of passionate political feuds, modern history was a dangerous field to teach. I completed my high-school studies in Zagreb in 1924.

I never doubted that I should study chemistry. I suspect that even at a young age, the strongest driving force for my interest in chemistry was my curiosity about an unknown, invisible world about which most lay people know practically nothing. My father, however, wanted me to choose a practical profession and not to become a scholar. Because there was no suitable place in Yugoslavia for chemical engineering studies, it was decided that I would study chemistry at the Czech Institute of Technology in Prague. One of the reasons for this decision was that my father had studied at Prague's Charles University and retained the most pleasant memories of that city.

Student of Chemistry in Prague (1924–1929)

In the fall of 1924, at the age of 18, I enrolled in the Chemical Engineering School of the Institute of Technology in Prague and thus separated myself geographically from my family. A totally new period in my life began. I quickly overcame the first difficulty, mastery of a new Slavic language, Czech. Even during my high-school years, I studied not only chemistry but also mathematics and physics; therefore, I more or less knew all the subject matter of the first semesters. Among the books I read in Zagreb, I specially treasured *The School of Chemistry* by Wilhelm Ostwald, which was translated into Croatian after the war. Thus stimulated, I studied, in Prague, Ostwald's other chemistry books and, later, also his philosophical writings. Later, I read also the works of Ernst Mach and Henri Poincaré, to which Ostwald referred. Between these demanding evening readings and the daily lectures and laboratory exercises, there lay a deep chasm. My readings dealt with large, general problems in the philosophy of science, such as the nature of space and time, whereas the lectures described endless details. I saw no way that I could move forward from those details to the general questions that I saw as the aim of science. An exception was the course in physical chemistry by Franz Wald, who lectured to undergraduates about his original ideas regarding the foundations of chemistry. In his Faraday Lecture,[2] Wilhelm Ostwald wrote about Wald as follows:

> Up to the present moment, the question whether it is possible to deduce the stoichiometric laws without the help of the atomic hypothesis has only been raised by other investigators in order to deny the possibility. So far as I am aware, there exists only one man who has worked upon the question with the earnest hope of obtaining an affirmative answer. The man is Franz Wald. . . . I feel it imperatively necessary to express my deep respect for and my thankful obligation to this solitary philosopher who has presented his work during a long series of years almost wholly without encouragement or sympathy from others.

As one of Wald's few listeners, I found out that he was, from my point of view, very old, 65 years. A few years later he died. Except for the examination with him, which I passed with distinction, I had no personal contact with him. Wald's ideas, which today could be designated as complementary to atomic and molecular chemistry, were not further developed. Nevertheless, the impressions made on

Rudolf Lukeš (1897–1960).

me by this creative thinker were unforgettable. I overcame my disappointment with the study of chemistry when I became acquainted with Rudolf Lukeš in the organic chemical laboratory. He conducted the *Praktikum* as assistant to the professor of organic chemistry, Emil Votoček. Until then, I had not appreciated organic chemistry at all. As presented in the second year, organic chemistry seemed to consist of endless details, compounds, and reactions without connection. Lukeš was only eight years my senior, and he had already begun to carry out independent research and needed a co-worker; thus I became his apprentice.

The laboratory for organic chemistry was administered by Emil Votoček, who was a student of Bernhardt Tollens. In this laboratory, the major work involved sugars, particularly methylpentoses. Lukeš, however, was fascinated by the bizarre-looking structures of alkaloids such as cocaine, quinine, and morphine, and had begun, self-taught, to work in this area. Throughout his life, he stayed true to the alkaloids and related heterocyclic compounds. External circumstances prevented him from attaining full development in this field. Until 1939, he was in a subordinate position as Votoček's assistant.

From 1939 to 1945, the Czech universities and institutes of technology were closed by the German occupation authorities. Those years were lost as far as scientific work was concerned. After 1946, when Lukeš became Votoček's successor, laboriously rebuilt the destroyed laboratory, and saw a lively scientific activity develop, he met with an early death in 1960. We were the best of friends and, throughout the whole period, stayed in the closest contact personally and through correspondence. I have more than once expressed my view that the best way to study science is as an apprentice to a master who is a model both in his field and in his personal characteristics. Through this collaboration, I learned that, in the beginning, it was important for a chemist to be confronted with reality and that sometimes it is better to follow the maxim, "Work now, understand later", rather than the reverse. Through my mentor, I first learned the systematics of organic chemistry and the organization of its literature, which allows one to survey the frightening number of known compounds and reactions and to push forward into the unknown. (Admittedly, there were only 400,000 organic compounds registered then; today there are more than 10 million.) In addition, I learned from Lukeš how to conduct chemical operations *lege artis*, from cork boring and glass blowing to organic elementary analysis.

Although microanalysis had already been introduced a few years earlier by Fritz Pregl, we still needed 0.1–0.3 g of material for each C, H, N, or halogen determination. We had to get up early to complete three to four determinations daily with two combustion ovens. The heat in the combustion rooms in the summer was unbearable. Practically all starting materials had to be prepared in relatively large amounts from the simplest and cheapest material, and this operation demanded much of our time. I can offer as an example the first research topic that Lukeš suggested to me in 1927, the reaction of N-methylsuccinimide with phenylmagnesium bromide.[3] For this reaction, I had to prepare succinic acid from ethanol through ethylene, 1,2-dibromoethane, and 1,2-dicyanoethane; methylamine from formaldehyde and ammonium chloride; and bromobenzene by the bromination of benzene. One of the products of the reaction I studied, N-methyl-2,5-diphenylpyrrole, formed magnificent crystals that showed not only a beautiful fluorescence but also strong triboluminescence, a phenomenon that I had not previously encountered.

The awareness that I had created a new substance, something that no hands had previously touched, gave me great pleasure, and I sought more such experiences. I therefore spent all my free time while a student assisting Lukeš in his research and published several papers with him during my student days. I finished my regular studies in the shortest possible time, in eight semesters, and passed the Diploma examination in June 1928 with distinction. The second half of my studies consisted of many descriptive technological subjects and analytical laboratory procedures that I mastered without difficulty but with little enthusiasm. I had to complete my studies as quickly as possible, because my father was compulsorily pensioned on political grounds and could no longer support me financially. Nevertheless, I wanted to complete the graduation project that I had already begun. At that time in Prague, the project could only be done under the aegis of a professor. I therefore requested, rather early, a subject from Votoček in the area of natural products chemistry but not in sugar chemistry, which I felt had stagnated after the brilliant advances made by Emil Fischer and his school. Votoček assigned me the problem of clarifying the constitution of the aglycone of a new glycoside, rhamnoconvolvuline, which had been isolated in his laboratory. I quickly determined that the aglycone is 3,12-dihydroxypalmitic acid.[4]

Even before I took my doctoral examination in June 1929, I had to look around for a job. Unemployment then was severe both in Czechoslovakia and in Yugoslavia, and I could not hope for employment in an institution that would allow me to devote my time to research. Fortunately I met and became familiar with a schoolmate of Lukeš, Gothard J. Dříza. He was a young entrepreneur who traded in chemicals and laboratory equipment and who planned to establish a laboratory for the preparation of chemicals that were not available commercially. He also wanted to carry out a doctoral project in that laboratory. Because he had asked me while I was still a doctoral student to design plans for such a laboratory and then to run it, I accepted his offer without hesitation, especially because I would have the opportunity of conducting independent research after official working hours.

Practicing Chemistry in Prague (1929–1934)

Dříza bought a house in Holešovice, a suburb of Prague. The house was remodeled and equipped, and I began to work immediately after receiving my doctorate. Part of the activities of the laboratory was the preparation of various chemicals that were not available commercially, such as ammonium sulfite for hairdressers, chloroacetophenone for tear-gas-producing ammunition for the police, and exotic items such as dimethyltelluronium diiodide for the army research laboratory, in addition to large quantities of standard solutions for titrations. My employment was semilegal, and officially I received no salary. The owner of the laboratory was equally unofficially my first doctoral student,[5] and Votoček was officially the supervisor. My position was delicate but by no means unpleasant, and it was a happy time for me. Dříza completed his doctoral examination with distinction and was very proud of acquiring this title despite his many other activities and duties.

Anyone beginning independent research must choose a topic. For me, the topic had to be important enough to justify spending evenings and nights in the laboratory. I caught Lukeš's enthusiasm for alkaloids, but I wanted to do something socially relevant (in plain words, something that could bring in some money). My first choice was the alkaloids of the *Cinchona* bark. Quinine was still the most important antimalarial drug. Its constitution had been known since 1908, but its configuration was not. Nevertheless, Paul Rabe had previously prepared dihydroquinine synthetically. For the quinuclidine part of quinine, no suitable starting material was available for large-scale technical production. Bicyclic amines resembling quinuclidine with nitrogen at the branching position had been little studied, and so I devoted myself to this problem and worked on its various aspects in Prague, later in Zagreb, and even in Zurich 20 years later.

In 1932, I had to serve for 9 months in the Royal Yugoslav Navy, which was not easy. However, through the navy, I became

As a sublieutenant in the Yugoslav Royal Navy in Dubrovnik in 1934.

acquainted with the beautiful Yugoslav Adriatic coast, and I matured personally in the harsh climate of military life. Through unaccustomed constant living together with a large group of individuals of various backgrounds and education, I learned tolerance. I spent the major part of my military service in the chemical laboratory of the marine arsenal south of Dubrovnik, where I had enough leisure to read the first four volumes of the main series of *Beilstein's Handbuch* in the library. While on military service, I fell ill with tropical malaria, and this illness only strengthened my intention to work on antimalarial compounds. On my return to Prague from military duty, I married Kamila Vítek in 1933. I met her in April 1927, when she was still a school girl in Prague. We have thus known each other for 63 years.

University of Zagreb (1935–1941)

In 1931, I received a letter from Ivan Marek, professor of organic chemistry in the Faculty of Technology of the University of Zagreb, asking me if I would like to succeed him when he retired in 1933. I enthusiastically assented. The negotiations, however, took a long time. Finally, to my disappointment, the faculty decided to fill the post in 1935 not with a professor but with a lecturer, a so-called university *docent*, that is, a teacher with all of the duties of a professor but with the compensation of a poorly paid assistant. My hope that my academic post would give me ample opportunity to devote myself to research failed, at first, to materialize. My predecessor had been mainly interested in the improvement of organic elementary analysis, and the laboratory was not equipped for synthetic work. Furthermore, the laboratory budget was minimal, and co-workers and technical personnel were lacking.

Unexpected help came from a small prospering pharmaceutical enterprise, Kaštel, Ltd., that produced mainly tablets, pills, and injectable solutions of known pharmaceuticals. When I came to Zagreb, one of its directors, Eugen Ladany, who was also part owner, decided to expand the scope of the firm by producing known, successful medicinal compounds that were not available commercially. At the same time, a research laboratory was to be established for the synthesis and pharmacological testing of new compounds. I was asked to cooperate, and I was offered personal financial support, as well as help for the university laboratory where I was to work also on medicinal chemistry. Although I knew that I would lose much time and some of my personal freedom, I accepted the offer without hesitation. Thus at this time, I served both as a university lecturer and as a consulting chemist. During the second world war, Kaštel, Ltd., was nationalized. After the war, under the name of Pliva, it became the largest Yugoslav pharmaceutical company with a sizable research institute.

Heterocyclic Compounds and Medicinal Chemistry

My first concern was to make the general plans materialize. Financial assistance enabled me to equip the university laboratory with the most urgent necessities and to take on doctoral students and hire co-workers. The future managers of the chemical and pharmacological research laboratories that were to be established at Kaštel were sent abroad for training with Ernst Fourneau in Paris and Franz Brücke in Vienna, and the necessary steps for construction were undertaken. The most difficult task was to decide quickly which known pharmaceuticals were to be produced. I suggested that an analog of Domagk's prontosil (rubrum) (Scheme 1), the sensational chemotherapeutic agent, should be made without causing patent conflicts between us and the then-almighty German chemical industry.

One of my first doctoral students, Dragutin Kohlbach, had prepared a series of sulfanilamide–azo dyes,[6] one of which had good properties. We decided to produce and market this dye. When the necessary construction work was starting for a production facility, a paper by Daniel Bovet and Filomena Nitti (later Mrs. Bovet) of the Pasteur Institute in Paris appeared in which these authors showed that the azo dye prontosil is reductively split in mammals to sulfanilamide and that this compound is actually the chemotherapeutic agent. Sulfanilamide had been known for a long time and was not covered by patents. It was not obtainable commercially, and we had already worked out the technical procedures for its preparation. Because the production facilities were in the process of construction, we were able to put the drug on the market under the name Streptazol in an extraordinarily short time. So long as only a small number of resistant bacterial strains existed, sulfanilamide was, despite some side effects, a wonder drug, particularly against the widespread cocci infections.

Not only the company but also the university laboratory and I personally profited from the resulting financial success. Because of the royalties, I could afford to spend several months in 1937 with

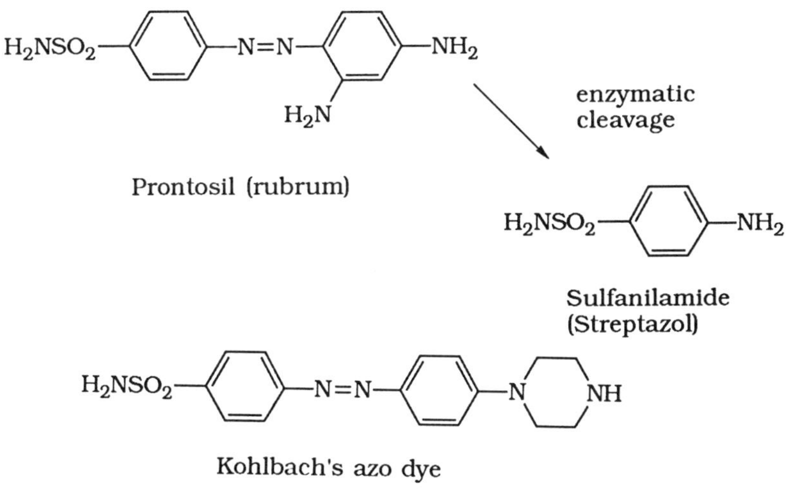

Prontosil (rubrum)

enzymatic cleavage

Sulfanilamide (Streptazol)

Kohlbach's azo dye

Scheme 1

Leopold Ruzicka in the organic chemistry laboratory of the Eidgenössische Technische Hochschule (ETH) in Zurich and to participate in 1938 in the major congress of the International Union of Chemical Societies in Rome. I thus broadened my scientific horizons and became personally acquainted with several leading chemists. Before arriving at the ETH, I did not know Ruzicka personally, but I chose to work with him because of his outstanding work and also because he was of Croatian origin, as I was. I hoped to learn how a famous chemist thinks and how his laboratory works. In Zurich, I worked on a triterpene, quinovic acid, and Ruzicka and I[7] published the results.

Neither the political atmosphere nor the scientific atmosphere in Zagreb prior to World War II was favorable for research; nevertheless, it was possible to inspire a number of students to engage in science. I imagine that these students, as unsuccessfully as I, sought to retreat through scientific work into an inner emigration to avoid facing the threatening reality of the political situation.

In Zagreb, the work that was begun in Prague on the synthesis of the *Cinchona* alkaloids was resumed. I would like to illustrate, with several examples, the ideas underlying this work and mention some of the expected and unexpected results. When we prepared the racemic diastereomeric 6-methoxyrubanols (1a) by following Paul Rabe's procedure for the synthesis of dihydroquinine, we discovered that these compounds were pharmaco-logically and chemotherapeutically similar to the corresponding natural alkaloids (1b).[8] The past master Rabe was not at all happy that we were poaching in his hunting grounds, especially because his doctoral student who had prepared the four stereoisomeric 6-methoxyrubanols had found only pharmacological activity but no chemotherapeutic activity. He wrote and told me of his displeasure. Rabe and our group finally agreed to publish the results in the same issue of *Chemische Berichte*.[9]

Later, we dared once more to invade Rabe's research territory. As an exercise, we made quinotoxine by starting with pure homomeroquinene that we had obtained by degradation of cinchonine and thus accomplished a partial synthesis of quinine.[10] Rabe did not seem to notice this invasion of his domain, probably because of the confusion of the war.

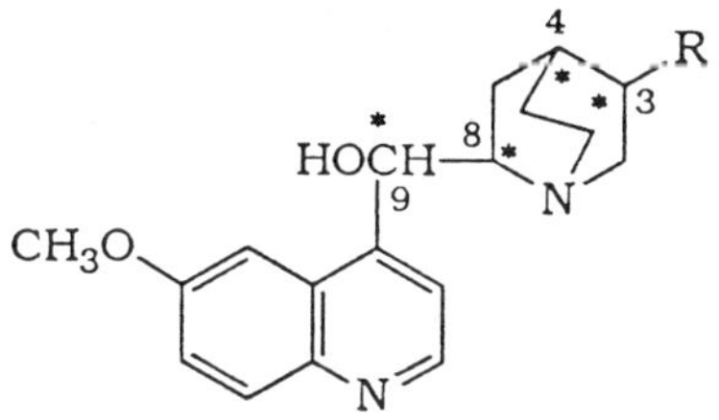

1a R = H

1b R = —CH=CH$_2$

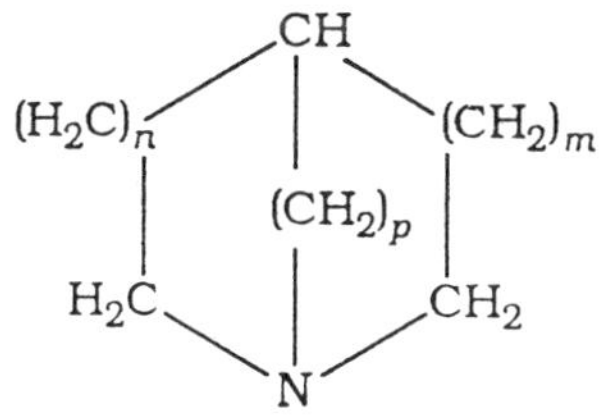

	2	3	4	5	6	7	8	9	10
$m+1$	2	2	3	3	5	5	4	4	3
$n+1$	2	2	2	3	5	3	4	3	3
p	2	1	1	1	0	0	0	0	0

The syntheses of quinuclidine itself (**2**)[11] and various homologous bicyclic tertiary bases (**3–10**)[12] with nitrogen as the branching atom were interesting in themselves because such structures occur in many alkaloids and had been little studied. In addition, the syntheses also led to three unexpected results that I particularly want to mention.

The synthesis of norlupinane, bicyclo[4.4.0]-1-azadecane (**8**)[13] drew our attention to George R. Clemo's postulated *cis–trans* stereoisomerism of norlupinanes. We guessed that in the so-called norlupinane B, which is formed by Clemmensen reduction of 5-oxonorlupinane (**11**), we were dealing not with a stereoisomer but rather with bicyclo[5.3.0]-1-azadecane (**7**) that is produced by rearrangement; we confirmed this guess by synthesis of bicyclo[5.3.0]-1-azadecane.[14]

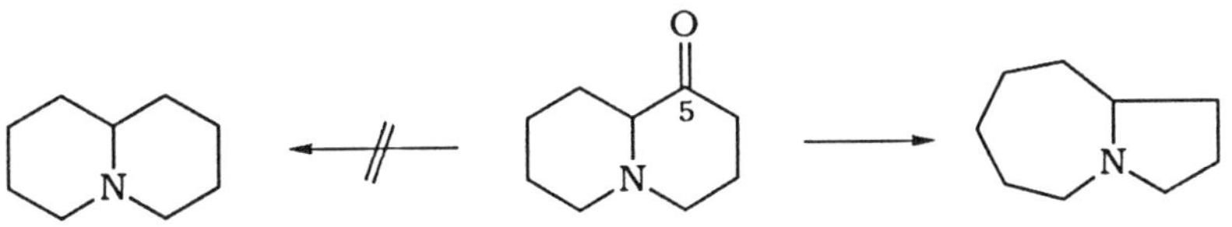

A further rearrangement was discovered when we tried to prepare β-collidine (3-ethyl-4-methylpyridine) (**12**),[15] which was needed for Rabe's synthesis of dihydroquinine, by dehydrogenation of **13**, which was easily prepared from acetone, formaldehyde, and methylamine. Instead of **12**, 2,3,4-trimethylpyridine (**14**) is formed through a novel type of rearrangement. Similarly 1-methyl-3-acetylpiperidine gives 2,3-dimethylpyridine.[16]

Finally we established that the γ- and δ-alkoxycarboxylic acid chlorides (**15**) required for further syntheses in this series very easily undergo intramolecular rearrangement to γ- and δ-chlorocarboxylic acid esters (**16**).[17] These unexpected observations confirmed my belief that organic chemistry was a magical world full of untapped possibilities and marvels that can be explored even with simple means.

The Adamantane Story

I will now describe the discovery of adamantane by Stanislav Landa,[18] which I witnessed while still in Prague, and our synthesis of this fascinating hydrocarbon later in Zagreb.

Although nearly half a century has passed, I still remember very clearly the impression left on my mentor, Lukeš, and on myself by Stanislav Landa's discovery of adamantane. At the Institute of Fuels of the Czech Institute of Technology in Prague, Landa was investigating the hydrocarbons of naphtha from Hodonin in Moravia, and he occasionally came to Lukeš for advice. One day, Landa told Lukeš that he had isolated a saturated terpene with the molecular formula $C_{10}H_{16}$ and melting point of 266 °C. Lukeš received this piece of news very skeptically.

One evening when I was with Lukeš, Landa asked us to come with him to see something really sensational. He was repeating the elemental analysis of his hydrocarbon. In the stream of oxygen in the combustion tube, his sample (weight > 100 mg) had sublimed, and we could see with the naked eye glittering tetrahedral crystals on the still cold wall of the glass tube. Lukeš immediately went to the blackboard and wrote the formula of "tetracarba-hexamethylen-tetramine".

Landa never mentioned this event in his publications. Instead he wrote that the structure was deduced on the basis of an X-ray Debye–Scherrer analysis.[18] Bizarre as it seems to me today, I decided to disclose Lukeš's divine intuition of the adamantane structure. This goal was the real reason that I started to work on adamantane synthesis, which was later accomplished in Zagreb. I described the event in our first publication. Unfortunately, the two papers on the synthesis of adamantane, which appeared in 1941,[19] are missing from many libraries. The lack of these papers may explain the statements found in the literature that we made only very small amounts of adamantane. Actually, by the second method starting with the Meerwein–Schürmann ester (**17**) through adamantane-dicarboxylic acid (**18**) and the reaction sequence COOH → Br → H, adamantane (**19**) was prepared in gram quantities.

We repeated the preparation in 1948 in Zurich for Peter W. Bridgeman, who wanted to heat adamantane under extremely high

17
Meerwein–Schürmann ester

18

19
adamantane

pressure to obtain diamonds. As Bridgeman wrote to me on March 1, 1948, "I have made, in all, five exposures to pressure. The results were, in all cases, negative." Later, when diamonds could be prepared by heating organic material at high pressures and temperatures, polyethylene was found to be just as good a starting material as adamantane. In the meantime, adamantane has become an easily obtainable compound, thanks to the elegant work of Paul von Ragué Schleyer.

Zurich Since 1942

When World War II broke out in 1939, we published the previously mentioned papers and several others, which were mainly about the syntheses of potential drugs, in German scientific journals. Through our publications in these journals, Richard Kuhn, who was then the president of the German Chemical Society, became aware of our work and invited me to give lectures in Germany. This invitation made it possible for me to emigrate to Switzerland in December 1941, after German troops occupied Zagreb earlier in the spring of that year. With Kuhn's letter of invitation in hand, I requested German officials in Zagreb to order the Independent State of Croatia to issue passports for me and my wife. Meanwhile, I informed Ruzicka of my situation, and he obtained Swiss entry visas for us. This emigration

PRELOG *My 132 Semesters of Chemistry Studies*

First day cover commemorating the 100th anniversary of the Czechoslovakian Chemical Society. The illustration on the stamp is that of a carbon skeleton of adamantane.

was by no means as simple as it appears from this account. We traveled in the beginning of December 1941 by train via Italy to Zurich and have lived in Zurich ever since. The lecture that I was to have given in Germany at that time was given in Heidelberg in the winter of 1946. On that occasion, I met the past master Karl Freudenberg, as well as Theo Wieland, Hans-Joachim Bielig, Günther O. Schenck, and several other German chemists who gathered around Richard Kuhn in the Kaiser Wilhelm Institute, which was half occupied by the American army.

A few weeks before I came to Zurich in December 1941, a rather large group of Ruzicka's older co-workers had left Switzerland en route to America, because as Jewish immigrants from Eastern Europe, they felt insecure in Zurich. Some of them later made noteworthy contributions to the development of the American pharmaceutical industry. Ruzicka's oldest assistant, Wolf Moses Goldberg, became research director of Hoffmann-LaRoche in Nutley, NJ; at Hoffman LaRoche, a second emigrant, Leo Sternbach, later discovered librium (chlordiazepoxide hydrochloride) and valium (diazepam), the psychoactive pharmaceuticals of the benzodiazepin series that fundamentally changed the treatment of the mentally ill. Under the guidance of a third emigrant, Georg Rosenkranz (later president of Syntex, Ltd. in Mexico), the contraceptive pill was developed and produced by that firm, another event that contributed greatly to the shaping of modern society. Of the research group leaders, only one remained in Zurich, Placidus Andreas Plattner, who is known for his proposal and subsequent proof of the azulene formula.

Organ Extracts

The vacuum created by the exodus of Ruzicka's competent and seasoned co-workers enabled me to perform immediately useful work in the Zurich laboratory. For several years before the war, with support from the Rockefeller Foundation, Ruzicka had asked the Wilson Laboratories in Chicago (adjacent to the slaughterhouses made famous by Upton Sinclair) to produce large quantities of lipids from various animal glands. He hoped that by using modern separation methods, particularly molecular distillation and the then-blossoming chromatography, novel biologically active materials might be isolated from the animal extracts. Tadeusz Reichstein's isolation of a host of hormones from the lipids of adrenal glands, served as the paradigm for this work. The investigation of large amounts of extracts from pig testes had already been started by Goldberg but without success. Thus, no one was jealous when Ruzicka entrusted its continuation to me.

I devoted myself to this problem with great enthusiasm, but the work did not lead to the spectacular results anticipated. I did succeed in isolating and identifying several crystalline compounds from the testicular extracts, but several of them were known derivatives of cholesterol that had been obtained earlier by Sune Bergström and Otto Wintersteiner through the air oxidation of cholesterol. Possibly, and even probably, we were dealing with artifacts.[20]

Nevertheless I isolated three compounds of some interest. Two of these compounds were unknown and exhibited a pronounced typical musk odor. They were identified as 3α- and 3β-hydroxy-16-androstene (20 and 21, respectively).[21] Their constitution was established by synthesis, and some other related steroids were prepared and tested for their olfactory properties.[22] Of these steroids, the ketone 22 corresponding to the two alcohols possessed a very intense odor.

Ruzicka was impressed with the relationship of these compounds with the male hormone androsterone (23) that had been isolated from urine and with the similarity of the structure of 22 with that of civetone (24), the olfactory principle of musk, which he had established decades ago. We found no use for this little discovery, but compound 20 was used later as a pheromone in pig breeding. Its presence was also shown in truffles,[23] a fact that explains why pigs,

20 $R^1 = H, R^2 = OH$
21 $R^1 = OH, R^2 = H$

22

even in antiquity, had been used in searching for this expensive delicacy that grows under a thick layer of soil. Advertisements in appropriate magazines also asserted that the odor of 20 makes men irresistible to women. In our work, however, we observed no such effect!

A third known steroid, which we were the first to isolate from a natural source, was 3-hydroxy-20-oxo-5-pregnene (25). Earlier, Hans Selye drew attention to the role of this compound in stress. Its isolation from pig testes also had a practical consequence. Compound 25 is a key substance in the industrial synthesis of sex hormones and, at that time, was protected in the United States by patents as a semisynthetic product. A major American firm, Glidden Company, challenged the validity of these patents, because, as a natural product, this substance could not be protected by patents in the United States and in most other countries. The isolation of 25 from testes was even repeated and published.[24] Anyway, by the time the courts finally made their ruling, the patent protection had run out.

Further studies of other organ extracts were even less successful than those with testes extracts and yielded only known steroids and compounds such as chimyl and batyl alcohols, which, for the first time, were found in mammals.[25] I was glad when Ruzicka no longer insisted that I continue or direct these studies. Only on one later occasion did I work with lipids of animal origin.

During the war, CIBA, Ltd. in Basel, which supported me financially, had produced extracts from the urine of pregnant mares in large quantities (several million liters) in order to isolate estrogenic hormones from them. In the process, large amounts of neutral nonphenolic lipids were also obtained. These lipids contain a complex mixture of C_{19} and C_{21} steroids, which Russell Marker studied in detail at Parke Davis in Detroit. In addition to the steroids and probably many other metabolites, many lighter volatile compounds were present. When we examined these compounds, we were surprised to discover that several were C_{13} compounds, which turned out to be derivatives of ionone or their rearrangement products.[26]

Examination of the compounds we isolated and whose structures we determined (26–33) showed that the compounds are very probably degradation products of carotenoids, whose relationship to these important natural products is similar to that of the C_{18}, C_{19}, and C_{21} urinary steroids to sterols. I have always regretted that these compounds have received so little attention from biochemists and biologists and that their origin and possible biological importance are still unknown. I hope one day they will be rediscovered. In any event, I left this field and have never returned to it.

In connection with our investigation of the compounds with the ionone skeleton, the pyridine base $C_{16}H_{25}N$ (34) caught my interest, because it also contains the ionone skeleton.[27] This base was isolated from Californian petroleum by American chemists. When I informed Ruzicka that we had synthesized the base, he told me that years earlier, in his laboratory, a small amount of a pyridine base $C_{16}H_{25}N$ had been isolated from musk. We showed by oxidation that this base, which we named muscopyridine, was a 2,6-disubstituted pyridine derivative.[28] Later, Georg Büchi at the Massachusetts Institute of Technology showed by synthesis that muscopyridine was a

α–Ionone

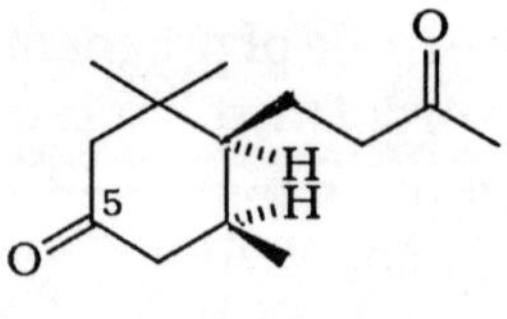

26

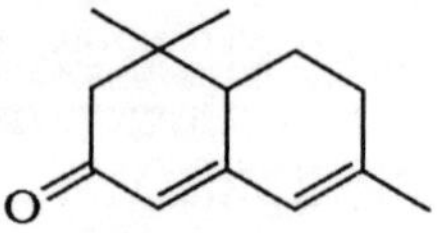

27

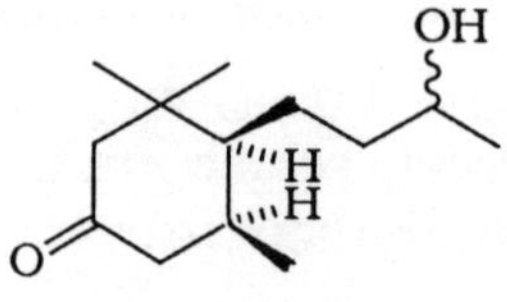

28

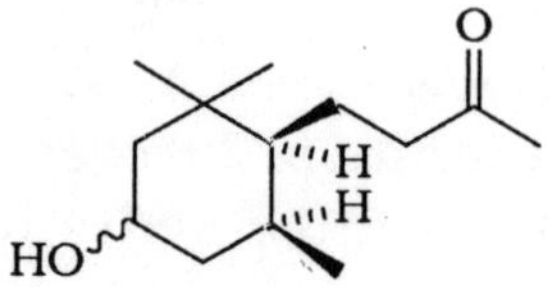

29

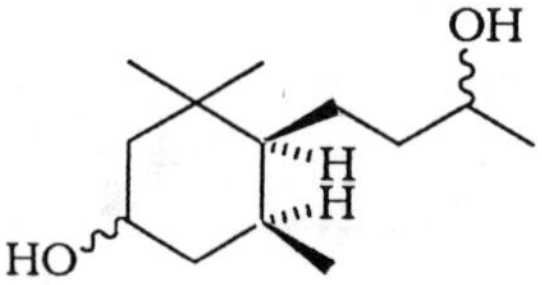

30, 31

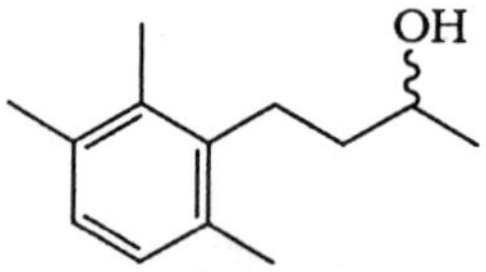

32

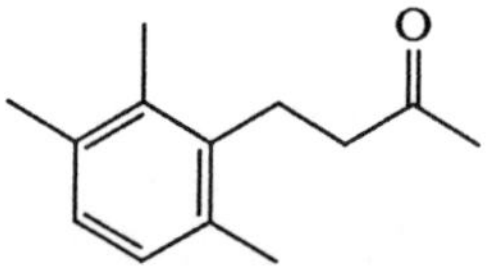

33

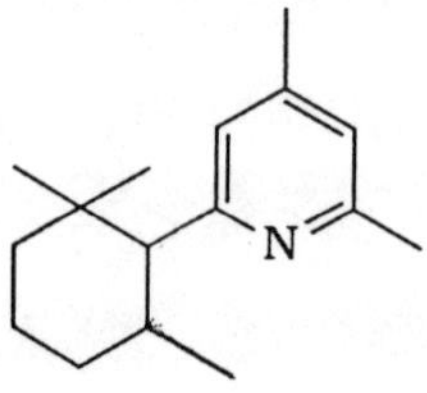

34

2,6-pyridinophane with the skeleton of muscone and thus quite different from the petroleum-derived base. For me, the investigation of organ extracts and other, more preparative studies not mentioned here that I did in connection with the steroid work under Ruzicka's auspices served as instructive exercises in the field of natural products, during which I particularly learned to work with small amounts of material. I learned many other lessons as well.

Alkaloids

My interest continued to focus on alkaloids, and Ruzicka knew it. Because Ruzicka in his later years had little interest in the chemistry of nitrogen compounds, he encouraged me to work independently on alkaloids with some doctoral students he assigned to me. Ruzicka was a fatherly friend and patron, not only to me but also to many of his other co-workers. He encouraged and helped me and sometimes even made decisions for me about my future. Together with Oskar Jeger, I have described Ruzicka's life and work in the *Biographical*

With Leopold Ruzicka (1887–1976) in Zurich in 1953. (Photograph courtesy J. D. Roberts.)

Memoirs of the Royal Society. His many-sided and dazzling personality would be impossible to describe within the framework of this volume.

During the war, cholesterol, the starting material for the preparation of steroid hormones, was difficult to obtain. Solanine from potato sprouts was one alternative. Its aglycone, solanidine, was revealed by dehydrogenation to contain a steroid skeleton. Our newly gained familiarity with the field of steroids, together with previous experience with the dehydrogenation of bicyclic bases with nitrogen as the branching atom and my predilection for alkaloids, found in solanidine an ideal research topic. On dehydrogenation of solanidine we obtained, in addition to a compound shown by George Soltys and Kurt Wallenfels to be γ-methylcyclopentenophenanthrene (35), a second product, 2-ethyl-5-methylpyridine (36), which led us to the correct constitution, 37, for solanidine.[29] A number of ring-A- and ring-B-modified solanidine derivatives that we had prepared behaved analogously to the corresponding cholesterol derivatives, and we deduced that the two series had the same configuration. This conclusion was later confirmed by others.

The choice of other alkaloids to work on was determined by a number of factors, some of them fortuitous. During the war, novel or rare natural products were not available, and thus I had to look for

Sir Robert and Lady Robinson the day after their wedding, Paris, 1957.

my research materials among inexpensive, easily accessible alkaloids. A further consequence of the war was that we were receiving very few professional publications, and of course, many of the scientific journals were not being published at all. Thus we had much time to read the older literature. When I discovered in the laboratory storeroom a considerable supply of strychnine, I decided to read carefully the many, often not easily understandable, publications of Hermann Leuchs (about 125 papers) and Robert Robinson (about 33 papers) on the structure of this alkaloid. After the war, I would probably never have found time for this diversion.

Because the strychnine formula suggested by Robinson in 1932 was not yet established as correct, I decided to follow the maxim of my doctoral mentor Votoček, "A newborn calf is not afraid of the tiger", and started to work on the structure of strychnine. I and Stefan Szpilfogel, who had previously worked on solanidine, were tremendously lucky in that after a few experiments we were certain that ring E in strychnine is not a five-membered ring, as required by Robinson's 1932 formula, **38a**. However, we missed our opportunity to deduce the correct formula when we proposed formula **38b**.[30] Robinson reacted to our suggestion as follows: "If Prelog and Szpilfogel are able to produce new evidence that ring E is six membered, the formula which must be considered is III."[31] Formula III corresponds to structure **38c**. This formula was confirmed by Woodward's brilliant analysis of all the known facts obtained through our experimental work and that of others, by Woodward's synthesis and, last but not least, by Bijvoet's X-ray structural analysis.

There is hardly another organic compound whose structure determination required as much experimental and intellectual work as that of strychnine. Our contribution cheered me, but of course I regretted my mistake. It was some consolation that before the constitution **38c** was finally accepted, Robinson, in a weak moment shortly after proposing the correct formula, published another unacceptable formula, **38d**,[32] which contained a quinuclidine skeleton. When we later discovered in cinchonamine the missing link between the *Cinchona* and the *Strychnos* alkaloids, Robinson stated, "I have sometimes made mistakes but never senseless ones." In 1944, we determined the still-unknown configuration (**39**) at C-3 and C-4 of the quinuclidine part of the major alkaloids of the *Cinchona* bark[33] and thus solved a problem of classical alkaloid chemistry.

Of other work from this period, the chromatographic resolution of racemic Tröger base (**40**) into its enantiomers should be mentioned. Tröger base was the first example of an optically active compound with trivalent nitrogen atoms as chiral centers.[34] This work, therefore, found its way into many textbooks.

The studies reported here were conducted or begun during the war. All of us in the laboratory, and particularly the immigrants, anxiously followed the course of the war and lived by the motto, "Work and don't despair." I myself slowly climbed the academic

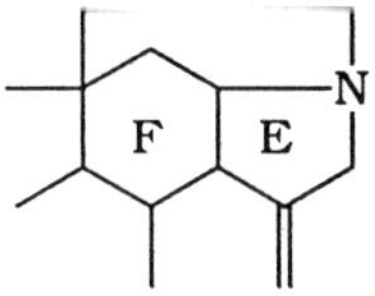

38a 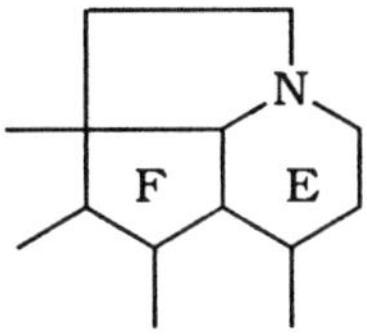 **38b** 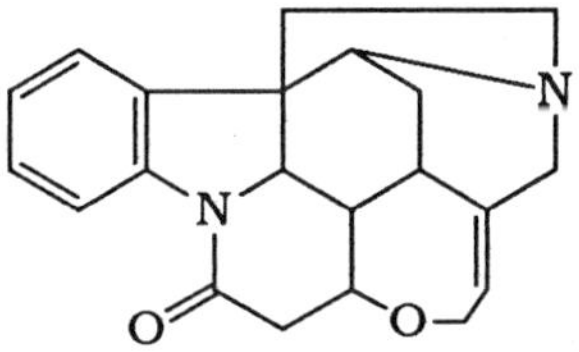 **38c**

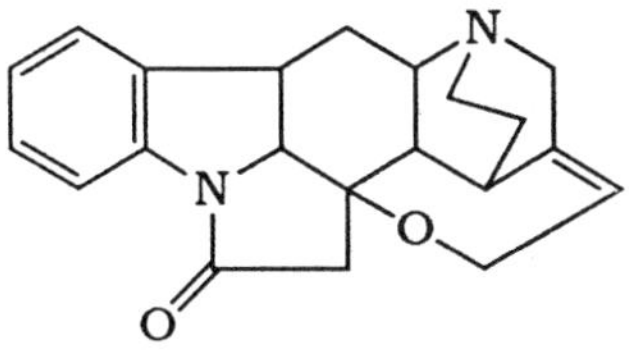

38d 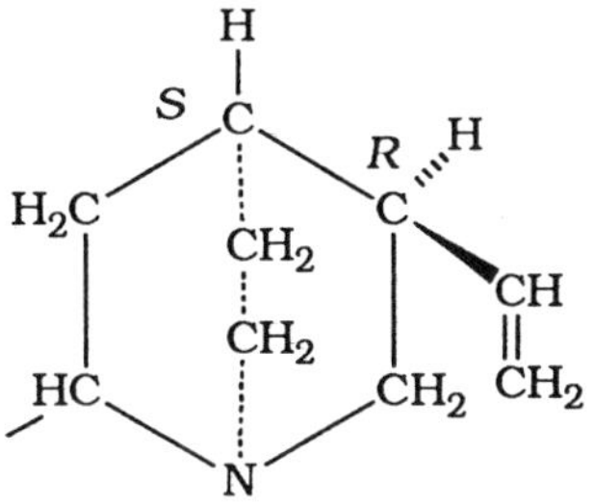 **39** (compare with formula **1**)

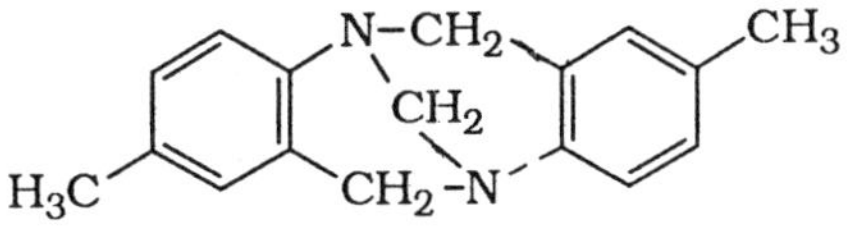

40

ladder. In 1942, I became *Privatdocent*, and in 1945, I was given the title, if not the salary, of a professor at the ETH. From 1943 on, I was financially supported by CIBA, Ltd. of Basel, which allowed my wife and me to live comfortably in a small apartment near the laboratory.

During the postwar years, 1945–1947, major changes were taking place not only in the world at large but also in our own work community in Zurich. Toward the end of the war and especially on its conclusion, the head of the laboratory, Ruzicka, devoted much time to organizing humanitarian aid to countries that had suffered from the war. To this activity was added after the war the establishment and expansion of his important collection of 17th-century Dutch paintings, which, for a time, occupied him completely. As a

Maurice Marie Janot (1903–1978).

consequence, his co-workers, who previously devoted practically all their time to Ruzicka's research problems, slowly broke up into groups working on their own topics. These topics deviated from those that were traditional in Zurich, partly because of the renewal of the lively contact with foreign chemists after the war. For European and American scientists, Switzerland, which was untouched by the war, was the favorite travel destination. Thus we became acquainted with many important chemists, among them Ruzicka's good friends and acquaintances, Roger Adams, James Conant, Louis Fieser, Maurice Marie Janot, Morris Kharash, Robert Robinson, and others.

My meeting with Janot led to a collaboration for many years that ended only in 1956. He had at his disposal a large number of exotic natural products of unknown structures from the French colonies. Most of the natural products belonged to the indole alkaloid group, and he suggested that we work on some of these materials together. His young colleague Robert Goutarel came to Zurich and spent some time with us; later we maintained contact by correspondence.

Of these joint projects, I want to mention in particular our determination of the constitution of corynantheine 41[35] and cinchonamine 42.[36] These constitutions revealed the close biogenetic connections between the many diverse indole alkaloids and the quinoline alkaloids of *Cinchona* bark. These biogenetic connections became the starting point for many studies by others in this field.

In other joint work with Janot and Goutarel we looked at sempervirine (1948), gelsemine (1951), pseudoyohimbine, yohimbine, corynanthine (1952), bankankosine (1952), ibogaine (1953), aricine (1954), and ibolutein (1956) but only obtained partial information, which in some cases we interpreted incorrectly.

In Zürich we were more successful with the aromatic *Erythrina* alkaloids, which, in contrast to other alkaloids we had worked on, we ourselves isolated from the seeds of *Erythrina abessinica*, Lam., a plant used in Zaire (then the Belgian Congo) as a shade tree on coffee plantations. Karl Folkers, the first chemist to work in this field, later supplied us with more material. Between 1950 and 1956, we determined, still by chemical means, the novel structures of erysodine (43) and erythraline (44)[37] and, by way of these, the structures of a number of other closely related alkaloids.[38] The aromatic *Erythrina* alkaloids, together with the related heteroalicyclic alkaloids studied in detail by Virgil Boekelheide at about the same time, form a closed group of natural products. When Boekelheide spent his sabbatical leave in Zurich, we wrote a review article about this group and other related indole alkaloids.[39]

41

42

43 $R_1 = R_2 = CH_3O$

44 $R_1, R_2 = CH_2\begin{smallmatrix}O-\\O-\end{smallmatrix}$

Collaborators

Let me now tell something about my co-workers who carried out the experimental studies and also significantly contributed to their design and interpretation. During the war, my co-workers were practically all undergraduate, diploma, or doctoral students of ETH. After the war, the situation changed rather markedly in that many visitors came from abroad and were with us for shorter or longer periods of time. Their names appear in the references.

Among the older and more experienced co-workers of Ruzicka with whom I collaborated, Oskar Jeger worked with me several times. In triterpene chemistry, mild conditions were necessary in place of the Bouveault–Blanc procedure as a general method for the reduction of a carboxyl group to a hydroxymethyl group. We suggested and tested the reduction of the thiol esters with Raney nickel.[40] But because the more convenient reduction of the carboxyl group with metal hydrides was discovered not much later, our method never gained much recognition. Another method we developed later for the synthesis of phenolcarboxylic acids from β-dicarbonyl compounds and acetone dicarboxylic esters[41] also found little use.

When R. H. Manske asked for a chapter on *Solanum* and *Veratrum* alkaloids for his monograph *The Alkaloids*, Oskar Jeger and I wrote the paper[42] together. This literary task made us work together experimentally on the *Veratrum* alkaloids, especially on the easily obtainable cevine. We soon discovered that the competition in this

At the CIBA Foundation, London, 1977, with Robert Burns Woodward (1917–1979). He typically smoked, drank coffee, listened carefully, and thought deeply. (Reprinted with permission from reference 111. Copyright 1986 Chimia.*)*

Sir Derek H. R. Barton.

field was frightening: Robert Burns (Bob) Woodward at Harvard and Derek Harold Richard Barton at Birkbeck College, London. We knew both of these prominent young organic chemists well.

Woodward had visited Zurich for the first time in 1948 when he had become known but not yet generally recognized in America through his quinine synthesis. A number of influential older chemists in the United States were slow to recognize Woodward's brilliance. They saw his early self-confidence as arrogance. Even on his first visit to Zurich, the way he planned, carried out, and interpreted his investigations and the unique way in which he described his results in his lectures impressed not only the younger members of our laboratory but even Ruzicka himself. The fact that we let Bob Woodward know how much we admired him gave him moral support, which he probably needed then. Until his so-untimely death in 1979, he remained a friend of our laboratory and my best personal friend. This friendship was a good start for scientific collaboration.

With Derek Barton, the initial situation was rather different. I first met him in 1947 at the first major postwar international chemical congress in London. It was an important occasion where I met many acquaintances and made many friends. Particularly, I met the young organic chemists working at Imperial College under the direction of Ian Heilbron. Derek was no longer with them but worked in the cellar of the laboratory for inorganic chemistry, on the foundation stones for his future fame. A former student from Zagreb brought me there, and our meeting led to a stimulating discussion. Later when Barton moved to Birkbeck College, his group and the Jeger–Ruzicka group were engaged in a vigorous contest in the field of tetracyclic triterpenes, each group hoping to reach the same goal first, as was usual in those days. In contrast, we believed that it was not worthwhile to compete in such cases, but rather, given sufficient trust, that we should communicate our results to each other. Unnecessary duplication, priority arguments, and disappointments are thus avoided, and one also learns much from one's partners.

Collaboration on the structure determination of cevine, which was considered a fairly difficult problem, was enjoyable and led to structure **45**,[43] in which the configuration of only one of the 13

45

The configuration of the chiral center in the box was false. The correct configuration was proposed by S. M. Kupchan (*Tetrahedron* **1959**, 7, 47).

asymmetric carbons was false. It is an irony of fate that this wrong configuration was assigned via conformational analysis. The cevine structure determination for which Jeger and his able co-workers on our side did the experimental work was also my last venture in the field of alkaloid chemistry. In what follows, I want to explain the personal and more general reasons for my decision to leave this area.

Medium-Sized Rings

One personal reason was that I began to be interested in topics not directly related to alkaloids. Placidus Plattner and I had to give lectures to advanced students, and we decided that we should discuss the electronic theories of reactivity that had not been dealt with in Ruzicka's introductory lectures. Ruzicka's comment was that he was not interested in theories that explained the regularities found in aromatic substitution but were unable to account for the large reactivity differences he observed in saturated alicyclic compounds. Shortly after the end of the war, Edgar Heilbronner urged me to attend a colloquium in the nearby laboratory for physical chemistry of the university directed by Hans von Halban in which the work of Kenneth S. Pitzer on free rotation around single C–C bonds was discussed. For me, it was the moment of enlightenment. I realized that here an answer could be found for many unsolved questions of aliphatic and alicyclic chemistry that Ruzicka could not explain with the Robinson–Ingold ideas. One of these questions was the strange behavior of simple ring compounds with 9- to 11-ring carbon atoms, in which I was then interested.

When we found that tetracyclic androstane derivatives **20–22** have an odor similar to those of the classic musk olfactory components civetone and muscone, I asked myself whether the corresponding bi- and tricyclic compounds also had similar odors. The compounds we wanted to prepare to test this hypothesis were cyclic compounds containing 9–11 carbon atoms; in those days, these compounds could be prepared only in very poor yield by the available ring closure methods. In a colloquium at ETH, one of Ruzicka's co-workers mentioned that V. L. Hansley had claimed in a Dupont patent that it was possible to obtain a 6.5% yield of 2-hydroxycyclodecanone from sebacic ester in hot xylene with molten sodium in a colloid mill. This yield was far more than had ever been achieved for a 10-ring cyclization.

Because we had no colloid mill, we tried the same reaction in a glass vessel with vigorous stirring and, to our surprise, obtained a 40% yield. We showed in 1947 that other many-membered rings could also be prepared in good to excellent yield by the acyloin synthesis (46 → 47), even those of medium ring size, without requiring very high dilution (Figure 1).[44] Our delight with this result was somewhat dampened when Ruzicka told us that at the laboratory of Firmenich & Company in Geneva, Max Stoll had obtained the same results but, to protect patent rights, had not published them.[45]

$$(CH_2)_n \begin{array}{c} \text{COOR} \\ \text{COOR} \end{array} \xrightarrow{Na} (CH_2)_n \begin{array}{c} \text{CO} \\ \text{CHOH} \end{array}$$

46   47

Under Ruzicka's pacifying influence, we finally agreed to publish the work on acyloin syntheses of many-membered ring compounds simultaneously. Fortunately the perfume industry was mainly interested in 15- to 17-membered ring compounds with musk odor. I, on the other hand, was fascinated by the unusual behavior of 9- to 11-membered rings that could now be prepared in this way. The "medium-ring effect" was evident not only in the difficulty of ring closure but also in abnormal physical properties and chemical reactivity, an anomaly already noticed in classic studies by Ruzicka.

Through the availability of a series of alicyclic and heteroalicyclic compounds with medium-sized rings, we demonstrated this effect in many new cases.[46] It was impressive, for example, that under identical reaction conditions cyclohexanone and cyclotetradecanone each smoothly gives cyanohydrins when allowed to react with hydrogen cyanide; conversely, cyclodecanone practically does not react (Figure 2).[47] We explained these results on the basis of models, showing that this behavior has to do with a constellational (conformational) effect. Alicyclic ring compounds

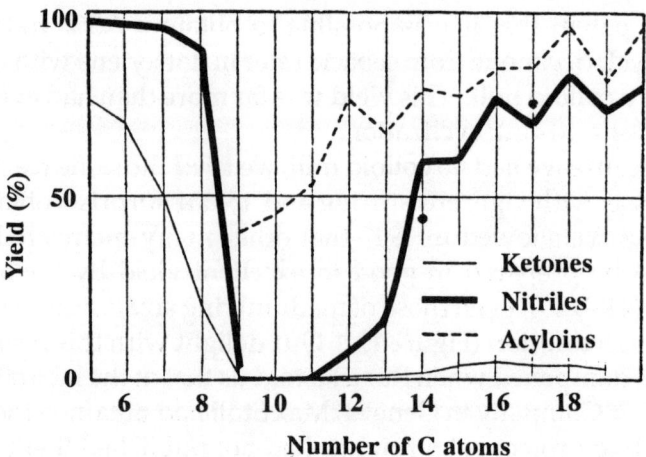

Figure 1. Yields of ring closure reactions vs. number of ring members. Key: —, pyrolysis of dicarboxylic acid salts (Ruzicka); ■, condensation of dinitriles under high dilution (Ziegler); and - - -, acyloin condensation from dicarboxylic acid esters (Copyright 1950. Reproduced with permission of the Royal Society of Chemistry.)

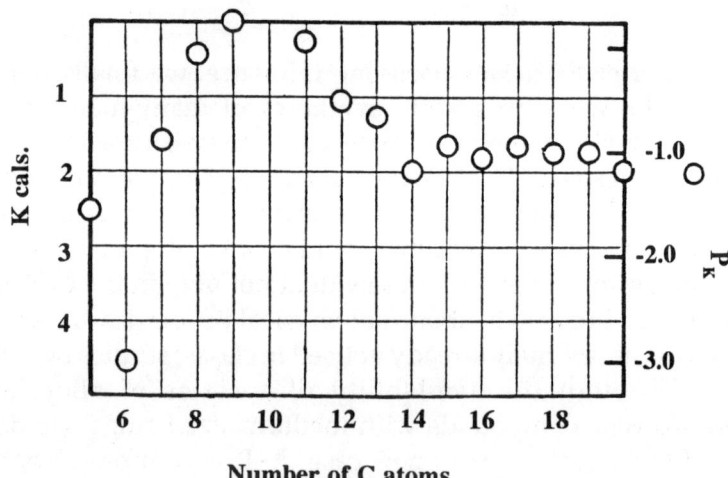

Figure 2. Negative logarithms of the dissociation constants and the free energy of the reaction cyclanone cyanohydrin ⇌ cyclanone + HCN are plotted against ring size, the influence of which is unexpectedly great. (Copyright 1950. Reproduced with permission of the Royal Society of Chemistry.)

exhibit nonclassical ("Pitzer") strain that depends on ring size and that changes with the nature of the transition state and products.

I surveyed these results in the area of many-membered ring compounds in 1949 in the first Centenary Lecture of the Chemical Society[48] at a number of places in England. The lectures were well received. The influence of ring size and ring form on the physical and chemical properties had been discussed much earlier, mainly in the case of five- and six-membered ring compounds, especially in natural products of the sugar, steroid, and terpene series. Walter N. Haworth did the pioneering work on sugars and also suggested the word *conformation* for various forms resulting from rotation of atoms around single bonds. In alicyclic chemistry, the chair and boat forms of cyclohexane had long been discussed. Following the German usage suggested by Friedrich Ebel, these forms were called *constellations*, the word I also used in my first publications. When Barton, after detailed inspection of extensive data on steroids and terpenes, established the field of conformational analysis, the term constellation disappeared from the literature.

The readily available many-membered acyloins were converted into various mono- and disubstituted derivatives, and the dependence of physical and chemical properties on ring size was investigated.

Attempts to prepare the bicyclic and tricyclic compounds with constitutions similar to those of the monocyclic musk odor chemicals and the tetracyclic steroids with musklike odor led to interesting observations on differences in reactivity between ordinary five- to seven-membered ring compounds compared with their large-ring homologs. According to Robinson, the reaction of the five- to seven-membered cyclanone-2-carboxylic acid esters **48** with Mannich bases gives the bicyclic α,β-unsaturated ketones **49**. When we carried out the reaction with the eight-membered ring ester, we obtained compound **50** as a second product. The higher ring homologs gave products only of the latter kind.[49]

This result was surprising, because the double bond in these compounds was in a bridgehead position. According to the well-known Bredt's rule, such compounds should not exist or, at least, should be unstable, whereas in our experiments, they were produced preferentially! Examination of models shows that even in a bridged bicyclic carbon skeleton with an eight-membered ring, a bridgehead double bond causes very little strain, and thus the validity of Bredt's rule is restricted to bicyclic compounds of normal ring size. It has since become known that with suitable forcing reactions, a double bond even at the bridgehead of a norbornane skeleton can be made, although such compounds are unstable.

Related to the question of the limits of validity of Bredt's rule was the question of the stability and reactivity of the metacyclophane derivatives 51–55. These derivatives are produced from cyclanones with larger than eight-membered rings, which condense readily with nitromalondialdehyde and smoothly undergo the subsequent reactions. From cyclooctanone, by contrast, compound 54 is obtained instead of 51. Despite the bridgehead double bond, 54 is apparently more stable than its corresponding aromatic counterpart.[50]

The substantial influence of ring size on the strain differences in the quinones 53 and the corresponding hydroquinone 55 (Figure 3) was measured polarographically by Karl Wiesner.[51] Wiesner came to Zurich as a Rockefeller Fellow from Jaroslav Heyrovský's laboratory in Prague in 1948 to escape from polarography and to devote himself to organic chemistry. After three years in Zurich, rather than return to Prague, he emigrated to Canada, where he founded a school in Fredericton, New Brunswick, that made major contributions to alkaloid chemistry.

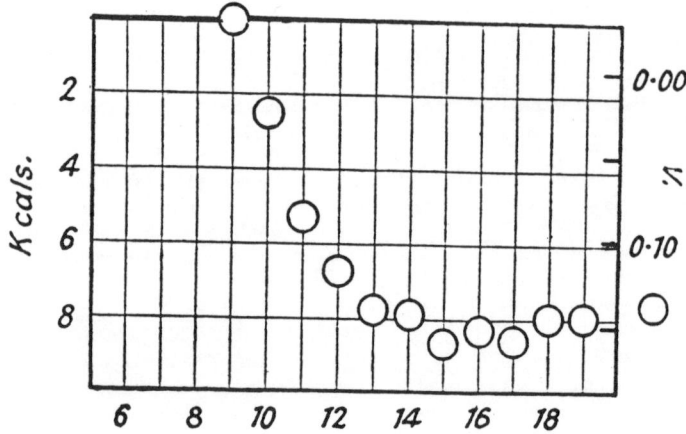

Figure 3. Free energy (in kilocalories per mole) and redox potential (in volts) of the equilibria 55 ⇌ 53 vs. number of ring members. (Copyright 1950. Reproduced with permission of the Royal Society of Chemistry.)

With Karel Wiesner (1919–1986) in Fredericton, New Brunswick, Canada, in 1972.

With Charles C. Price, head of the Department of Chemistry at Notre Dame, March 5, 1950. Vladimir Prelog was the Reilley Lecturer in Chemistry at Notre Dame.

Lectureship at Notre Dame (1950)

At this point, I want to interrupt my account of the chemical work done or started between the end of the war and 1950 and describe my career and the great changes in style and methodology of organic chemistry since 1950. As life after the war became more or less normal, there was a great need to become acquainted with chemists and developments in chemistry in other countries. We in Switzerland wanted to meet chemists from the United States particularly. When Charles C. Price invited me in 1950 to spend several weeks as Reilly Lecturer at the University of Notre Dame in South Bend, Indiana, I was not able to resist the temptation, although the time was not convenient because my son Jan was born in May 1949 and it was painful to leave the family.

The crossing on the *Queen Mary*, on which Edgar Heilbronner and his wife were also traveling on their way to Pasadena to work with Linus Pauling, was very stormy. We arrived in New York two days late. Flight connections were suspended because of bad weather, and I traveled in a packed overnight train to Chicago and South Bend. Kenneth Campbell and Ernest Eliel met me at the station, took me to dinner, and then escorted me to the university, where I gave my first lecture in my very limited English. In other ways, too, my stay in the United States was hectic because I wanted to see and experience as much as possible. I gave my two lectures at the University of Notre Dame in the middle of the week. At the end of that week and the beginning of the next, I traveled in the Midwest, visiting Chicago, Evanston, Lafayette, Kalamazoo, Ann Arbor, Indianapolis, Urbana, Minneapolis, and Madison. To these places should be added Cambridge (where I visited Woodward), Philadelphia, Washington, and New York on my return trip. All in all, I gave 25 lectures. I was deeply impressed by the friendliness of the many colleagues I met during my expeditions and the openness with which they discussed their unpublished or planned work.

In South Bend, I lived first in a hotel for three weeks and then in Price's home and was taken into his family. At that time, as well as later, he was deeply engaged in politics and was running for senator from Indiana. He spent many evenings talking to potential supporters (mainly members of trade unions) to inspire them with his ideas for world government. Nevertheless, he failed to win the nomination. I sometimes joined him and thus became acquainted not only with American academic communities but also with the working class of America. After two months in the United States, on the calmer return voyage on the *Queen Elizabeth*, I was able to reflect on my many impressions and experiences and to ask myself if I could live in America and be happy there.

Full Professor in Zurich, Lectureship at Columbia University (1951): Revolution in Instrumentation and Other Developments (1950–1957)

On my return, I had to make important decisions. Harvard University had offered me a well-paid research professorship, and Ruzicka, whose advice I sought, decided that the ETH must either promote me to a full professorship or I would have to go. Under his strong urging, the ETH president, Hans Pallmann, created a personal chair for me, something unusual at the time. The situation was delicate, because my older colleague Plattner, who in 1949 was appointed associate professor together with me, could not be promoted to full professor at the same time. It was customary then that an institute had only one full professor; having two was already a rare exception. Plattner had made major contributions not only as researcher but also as teacher, organizer, and administrator and was deeply hurt by this discrimination. This circumstance played a major role in the events of the following year when I was again in the United States for an extended period.

In 1951, the American Chemical Society celebrated the festive 75th anniversary (diamond jubilee) of its founding. The celebration was combined with an IUPAC (International Union of Pure and Applied Chemistry) Congress in New York and was attended by many participants from all over the world. I crossed the Atlantic this time with my wife, and on the *Caronia* were many chemists, among them William Dauben (returning from a stay in our laboratory), E. R. H. (Tim) Jones, Charles Shoppee, and Alexander Todd and their wives.

To my surprise, during the congress in New York, Max Furter, director of Hoffmann-La Roche, invited me to accept the post of research director of his firm. I knew I was not suited for the position, and I declined this financially generous offer without much thought. Plattner then became research director and later one of the general directors of Hoffmann-La Roche in Basel. He contributed much to the development of research in this firm and in Switzerland.

After the congress, I traveled with my wife by train to the west coast, via Washington and Chicago, and visited San Francisco, Palo Alto, Los Angeles, and Pasadena. I met Donald J. Cram for the

Placidus Andreas Plattner (1904–1975) in Zurich.

first time in Los Angeles and discovered our common interest in asymmetric synthesis. Our return journey, with the Grand Canyon as an intervening stop, ended in New York, where I spent about two months at Columbia University as the Falk–Plaut Lecturer.

The lectureship at Columbia University was a unique experience for me. The head of the department of chemistry was Louis Hammett, whom I respected greatly. My host was William von E. Doering, who besides his post at Columbia, also directed a private laboratory in Katoonah, near New York. Other members of the faculty at Columbia were older chemists such as Robert Elderfield and younger ones such as Harold Conroy and Fausto Ramirez, with whom I maintained close contact. New York at the time was a much friendlier city than it is today, and the subway was a safe means of transportation; one could take a walk at midnight in most parts of the

With Kamila Prelog in Cambridge, Massachusetts, in 1951. (Photograph courtesy J. D. Roberts.)

city without fear. I smoothly fitted into the Columbia program and took part in lectures, colloquia, and even examinations. In addition, Randolph Major, research director at Merck, invited me to lecture once a week at Rahway in the evening. There I came into close contact with the superb researchers of this firm, including Karl Folkers, Stanton A. Harris, and Louis Sarett. In New York, we lived in an apartment with a view of the Hudson River and the Palisades and enjoyed the marvelous autumn weather. Nevertheless, I did not regret my decision to stay in Zurich.

After our stormy return voyage on the *Ile de France*, I had a number of problems to solve. Ruzicka and his first wife were divorced after a long marriage, and he married again in 1952. He found his way back to chemistry, but his interest was now centered less on day-to-day matters than on general considerations regarding the biochemical and philosophical implications of his earlier research. Further contributing to this change of Ruzicka's focus were changes in the methodology of organic chemistry, particularly in the field of structure determination, which he followed with interest but no longer mastered. After Plattner's departure, Ruzicka transferred the administration of the laboratory to Bruno Engel, who carried it out with insight and skill. In 1950 Ruzicka had entrusted the main lecture course on organic chemistry to me and attended it from 1952 to 1953 regularly every morning. He stimulated a number of former students and later group leaders to carry out independent research and to complete their *Habilitation*, among them Emil Hardegger, Oskar Jeger, Andor Fürst, Hans Heusser, Hans Günthard, and Edgar Heilbronner. Günthard was promoted to associate professor of physical organic chemistry in 1952, left us in 1958, and was made head of the ETH laboratory of physical chemistry in 1959. Fürst and Heusser later followed Plattner to Basel; the others remained in our laboratory.

As mentioned earlier, the methodology and style of experimental organic chemistry changed considerably after 1950. Physical methods, X-ray structure analysis, and molecular spectroscopy largely displaced the chemical methods that had been used almost exclusively earlier. It now became possible in a short time and with small quantities to determine structures more reliably and confidently than had been possible by chemical means. Furthermore, the isolation methods, particularly chromatographic procedures, were developed and refined to such an extent that reaction and natural product mixtures that formerly had been considered hopeless could be effortlessly separated into their components.

Other developments were occurring: the invention of efficient selective preparative procedures that permit rational planning and execution of syntheses of specific stereoisomers and multifunctional molecules; the creation of new physics-based concepts and models for the theoretical (especially mathematical) handling of molecules and their interactions; expansion of the organic chemist's repertoire by making use of atoms not used previously; and numerous successful advances in linking living processes with molecular events. These advances, and especially the so-called instrumental revolution, had a number of important consequences. The field of organic chemistry was enormously broadened, but the research costs also rose slowly by one or two orders of magnitude. The new methodology not only required expensive equipment such as diffractometers and infrared, Raman, nuclear magnetic resonance, electron spin resonance, and mass spectrometers and their associated computers but also specialists to look after them.

A further consequence of this development was that structure determination, until then considered one of the most important and intellectually demanding tasks, had now become largely routine. X-ray structure analysis was taken totally out of the hands of the organic chemists. The times had passed when structure determination demanded a profound knowledge of the reactivity of organic compounds, an ability to discern relationships among disparate data, as well as imagination and an instinct often based on presumed but not yet established biogenetic relationships, such as Ruzicka's isoprene rule. Many excellent chemists therefore turned away from structure determination and devoted themselves to synthesis or to the emerging field of theoretical chemistry.

My colleagues and I were aware of the significance of these rapid changes and, with Ruzicka's full support, undertook to build up the costly infrastructure demanded by the new style. Günthard, and later Heilbronner and his student Wilhelm Simon, contributed greatly in setting up molecular spectroscopy and instrumental analysis. For X-ray structure analysis, Jack Dunitz joined us in 1957, and for mass spectrometry, Josef Seibl came in 1960.

Microbial Metabolites and Antibiotics

My research group remained faithful to the chemistry of natural products, and we welcomed the new possibilities for structure determination. During the war, Plattner and his research group worked closely with the Institute for Special Botany of the ETH, which was directed by the phytopathologist Ernst Gäumann. They first isolated and investigated the microbial products that botanists had shown to cause the wilting of plants. Later, with encouragement from CIBA, they broadened this collaboration to include other microbial metabolites, particularly those with antibiotic activity, which the Basel firm tested for chemotherapeutic and pharmacological activity. When Plattner left Zurich before antibiotic research had really developed, I took over the field with the help of Plattner's student Walter Keller.

I have always enjoyed changing and expanding my field of activity. The motivation for changing research direction often involved novel types of structures with intriguing biological properties that were also of intense interest to our industrial friends and sponsors. Very soon I realized that microbial cultures were a gold mine of biologically active compounds with unusual structures. I therefore decided to give up completely my work on alkaloids, steroids, and terpenes of plant and animal origin and to devote myself to these novel natural products. I was involved with structure determination of the microbial metabolites **56–71** that were isolated from cultures of *Streptomyces* strains. After 1970, I left this field; it was taken over with great success by Keller.

A few general, as well as special, comments must be made about the structure and properties of the microbial metabolites we worked on. We were surprised to find that most of the substances studied contained many-membered rings. Many-membered ring compounds had been discovered at the ETH by Ruzicka during his investigations on musk odors. Ruzicka also prepared and investigated numerous simple many-membered ring compounds that were unknown at that time. They were considered rarities, and yet they occur very often in microbial metabolites. The first new compound we isolated in this series was a 14-membered lactone, the macrolide narbomycin (**56**). By oxidative degradation, we obtained compound **57**, called the Prelog–Djerassi lactone by other investigators.[52] This lactone carboxylic acid plays an important role as an intermediate in

56 Narbomycin

57
1,5-lactone of (2R,4S,6S)-trimethyl-(5S)-hydroxyheptane-1,7-dioic acid

the synthesis of macrolides and has been prepared synthetically more than 30 times since 1975. Because of its rather cumbersome systematic name, it was easier to call it by the given trivial name. My name became better known for this lactone than for any other accomplishment of mine.

Among the 32-membered macrotetrolides, nonactin (58a) was of particular interest because of its *meso* configuration and achirality, despite its 16 stereogenic carbon atoms.[53] Beyond that, the macrotetrolides 58a–d are ion-selective ionophores that played a role in the development of ion-selective electrodes by Simon in our laboratory.[54] The many-membered cyclopeptide echinomycin (63), whose structure we established almost correctly,[55] later became well known when it was found that it forms interesting complexes with nucleic acids.

Particularly rich in consequences was the discovery of ferrioxamines by Hans Zähner in the Institute for Special Botany. First, iron-containing compounds with antibiotic activity, ferrimycins, were found in microbial cultures. During attempts to isolate them, the activity disappeared, only to reappear after several purification

	R_1	R_2	R_3	R_4
58a	CH_3	CH_3	CH_3	CH_3
58b	CH_3	CH_3	CH_3	C_2H_5
58c	CH_3	C_2H_5	CH_3	C_2H_5
58d	CH_3	C_2H_5	C_2H_5	C_2H_5

Macrotetrolides: **58a**, nonactin; **58b**, monactin; **58c**, dinactin; **58d**, trinactin.

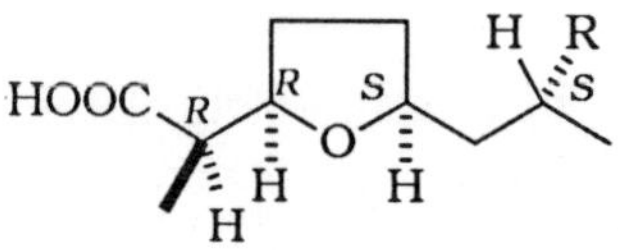

59a (−)-Nonactic acid, R = CH$_3$

59b (−)-Homononactic acid, R = C$_2$H$_5$

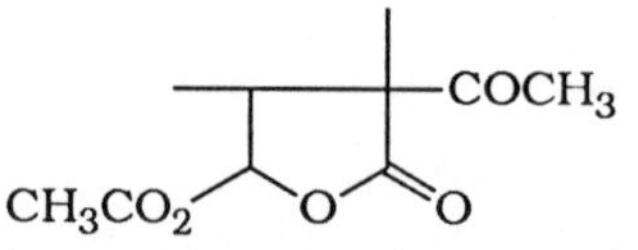

60 Acetomycin

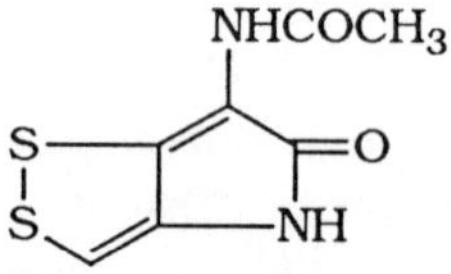

61 Holomycin

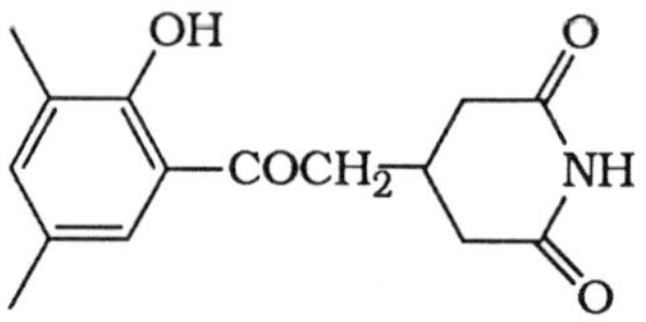

62 Actiphenol

steps. These puzzling results finally led to the conclusion that these antibiotics occur mixed with their closely related iron-containing antagonists, which inhibit their antibiotic activity. When we separated these antibiotic–antagonist mixtures, we determined the structures of both the antagonists and the antibiotics. We found that the antagonists are trihydroxamic acid–iron(III) complexes **64–67**. The antibiotic ferrimycins, such as **68**, are related structurally to their antagonists and even under mild conditions are converted into these, a situation making the biological proof during their isolation even more difficult. The ferrioxamines gained medical significance when the hematologists L. Heilmeyer and F. Wöhler in Freiburg im Breisgau found that their iron-free components, the desferrioxamines, were useful as medicines. They found that "pathological iron," which in certain progressive illnesses is deposited in various organs, is complexed by desferrioxamines and is thus eliminated in the urine. The desferrioxamines differ from other known complexing agents such as EDTA in that they bind iron(III) and aluminum(III) very strongly, whereas their bonding of other biologically important ions such as calcium(II), magnesium(II), and zinc(II) is weak so that the body tolerates them well.[56]

The boxed area of the structure was corrected in 1975 by D. J. Williams as follows:

63 Echinomycin

Finally, boromycin (**69**), the first boron-containing organic natural product, was discovered, and its structure was determined by X-ray analysis.[57] This compound is an orthoboric ester of a macrodiolide with a 32-membered ring.

An important group of antibiotics discovered not in Zurich but by researchers at Lepetit Ltd. in Milan and made available to us by Piero Sensi were the rifamycins. Together with the doctoral student Wolfgang Oppolzer, we were able to determine the structure of compounds **70–73** in competition with the X-ray structure analysts.[58] The rifamycins were later modified by research chemists at Lepetit and Ciba–Geigy. The drugs thus obtained, such as rifampicin, play an important role in combating tuberculosis and certain other illnesses, such as leprosy.

A deep blue-red antibiotic of plant origin, biflorin, was made available to us by O. Gonçalves de Lima of Brazil. In 1963, our chemical results and the nuclear magnetic resonance studies carried

64 Boromycin

65 Ferrioxamine B $R_1 = CH_3$, $R_2 = H$
66 Ferrioxamine D $R_1 = CH_3$, $R_2 = CH_3CO$
67 Ferrioxamine G $R_1 = (CH_2)_2COOH$, $R_2 = H$

68 Ferrimycin $R_1 = CH_3$, $R_2 = $ HO—⟨⟩—CO–, NH($C_9H_{16}N_3O_4$)

69 Ferrioxamine E
(Fe^{3+} complex of nocardamine)

out by Lloyd M. Jackman led to the unusual diterpenoid 1-oxa-phenalenequinone structure (**74**), which accounts for the deep color.[59]

Transannular Reactions and Structure of Medium-Sized Rings

After 1950 and concurrent with the work on natural products, the investigations begun earlier on many-membered ring compounds were continued, expanded, and refined. A special event in the field of many-membered rings was the discovery of transannular reactions and the related nonclassical course of substitution and elimination reactions in medium-sized ring compounds. We observed transannular reactions for the first time in 1951, when we oxidized the stereoisomeric cyclodecenes with performic acid. From *trans*-cyclodecene (**75**), we obtained, among other products, 1,6-cyclodecanediol (**77**), 6-hydroxycyclodecene (**78**), and *trans*-decan-1-ol (**79**).[60]

The formation of these unusual reaction products could be explained by assuming that the primary reaction product, the epoxide (**76**), forms a hydroxycarbonium ion under strongly acid conditions. Then the hydroxycarbonium ion yields the stable reaction products by a 1,5-hydride shift. At the same time, Arthur C. Cope observed similar reactions with cyclooctenes,[61] and we demonstrated analogous transannular reactions in cyclononenes[62] and cycloundecenes.[63]

70

71

72

73

74

However, cyclododecenes[64] again behaved classically in that they were oxidized under the same conditions to 1,2-diols.

I met Cope in 1951 after the ACS Diamond Jubilee Meeting in New York, and we exchanged information. We agreed that he would study transannular reactions with eight-membered ring compounds and we would investigate larger rings. He was not too happy about this arrangement, and the relationship was rather cool.

The assumption of a carbonium ion intermediate during the unexpected course of cycloalkene oxidation with performic acid led us to investigate the reactions of monosubstituted cycloalkanes in which carbonium ions are assumed as intermediates or transition states, such as the acetolysis of sulfonic esters of cyclanols or the reaction of primary cycloalkylamines with nitrous acid.[65]

The rates of acetolysis of homologous cycloalkyl *p*-toluenesulfonates were measured in our laboratory and that of H. C. Brown, and the results agreed in showing that compounds with medium-sized rings undergo much more rapid acetolysis than do their smaller and larger ring homologs; for example, cyclodecyl derivatives react some 300 times faster than cyclohexyl derivatives.[66] After alkaline hydrolysis, cyclanols and cycloalkenes were isolated as products of acetolysis. By labeling the starting material with ^{14}C and ^{2}H and by systematic degradation of these products, we found that both the substitution and elimination reactions in 9- to 11-membered compounds must have been accompanied by transannular nonclassical hydride shifts, with the substitution or elimination taking place partly in the C-5 or C-6 position.[67]

The conjecture that this abnormal behavior of medium-sized ring compounds depends on nonclassical strain was confirmed by X-ray analyses carried out by Jack Dunitz and his co-workers in our laboratory and correlated with other cases. The structure of the cyclodecane framework deduced from structural analysis of several of its derivatives may serve as an example (structure **A**). Its C–C–C

A

bond angles are spread markedly (classical Baeyer strain), the torsion angles deviate substantially from their optimal values (nonclassical Pitzer strain), and the distances between the intraannular hydrogen atoms are so short (1.95 Å by neutron diffraction) that they contribute significantly to the strain.[68] By distributing the total strain in medium-sized ring compounds over these three components—a fascinating and intriguing mathematical problem at the time—the alternative energetically preferred structures can be calculated by using the force field method. The experimentally determined structures were thus ideal objects for testing the efficacy of the various methods of molecular mechanics, as has been shown, among others, by Lifson and Dunitz ("Anybody can make a force field for alkanes").[69]

Beyond this application, the elucidated structures provide a solid foundation for mechanistic theories regarding the causes of the remarkable reaction behavior of medium-sized ring compounds:

1. The strain is diminished when the intraannular hydrogen atoms in reaction products or transition states are removed,

for example, when tetrahedral framework atoms become trigonal.

2. The positions of intraannular hydrogens favor the otherwise rare transannular hydride shifts.

3. Geminal substitution in the framework strongly influences the conformation and thereby the reactivity of medium-sized ring compounds, because there is no room to replace the intraannular hydrogen atoms by larger substituents without considerably increasing the strain.[70] Conformations must change instead.

I have not mentioned the dehydrocyclizations of many-membered ring compounds with palladized carbon.[71] These reactions proceed relatively smoothly if the cycloalkane or cycloalkene framework can aromatize without rearrangement. Such is the case with starting materials having $4n + 2$ ring members, that is, in 10-, 14-, or 18-membered rings. Cyclodecane and cyclodecenes under kinetic control do not give stable naphthalene predominantly but rather give up to 80% of azulene (80).[72] This reaction turned out to be one of the simplest syntheses of azulene. Cyclotetradecane is converted to a mixture of anthracene and phenanthrene, and cyclooctadecane is converted to triphenylene. Cycloalkanes with $4n + 1$ ring members give aromatic hydrocarbons containing a five-membered ring. Cyclononane yields indene, cyclotridecane fluorene, and cycloheptadecane 1,2-benzfluorene. If the cyclic hydrocarbon contains $4n$ ring members, skeletal rearrangements occur. Cyclododecane forms acenaphthene (81), and cyclohexadecane forms fluoranthene (82). Cycloalkanes with $4n + 3$ ring members such as cycloundecane and cyclopentadecane dehydrocyclize only with difficulty and yield complex product mixtures.

S. Kaarsemaker and J. Coops in 1952 and H. van Kamp in 1960 prepared by our method very carefully purified many-membered cycloalkanes and investigated them thermochemically. They found that hydrocarbons with 8- to 11-membered rings show a remarkable thermochemical ring strain, compared with cyclohexane, with a maximum in cyclononane and cyclodecane.[73]

Since 1968, when our last paper on medium-sized ring compounds appeared, this field has received less attention than I had

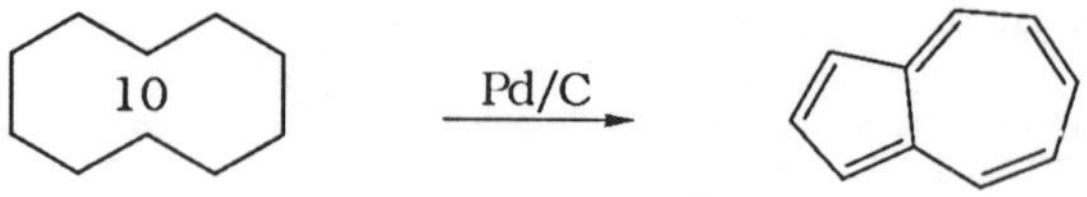

10 →(Pd/C) 80

12 →(Pd/C) 81

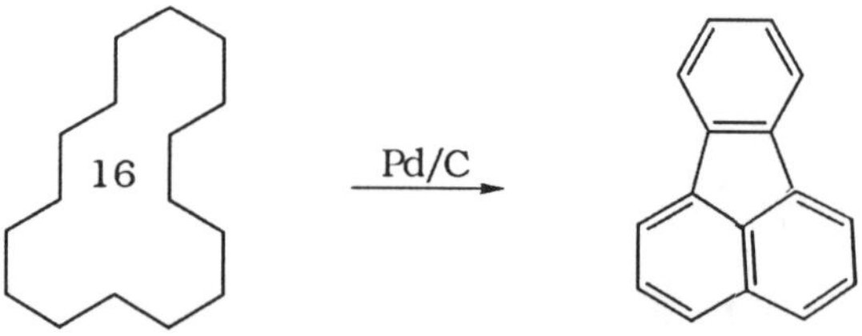

16 →(Pd/C) 82

hoped. One reason might be that many-membered ring compounds are barely or only marginally mentioned in most textbooks.

Asymmetric Reactions

Another field in which we began work in 1950 was that of *asymmetric synthesis*. I became aware of this field through a review by Alexander McKenzie in 1936, in which he mentioned, among others, numerous cases of *asymmetric induction* that he had observed in the reaction of α-ketocarboxylic esters of enantiomeric alcohols with Grignard reagents. After saponification of the reaction products, he obtained optically active α-hydroxycarboxylic acids.[74] However, despite his extensive data, the relationship between the configurations of the chiral alcohol and the resulting α-hydroxy acid formed in excess was not apparent. In the meantime, ideas about the energetically preferred conformations became clearer, and the relative and absolute configurations of starting materials and products were determined.

Thus it was opportune to reconsider the steric course of such reactions. On the basis of data collected by McKenzie (and some experimental corrections by us), a general rule was formulated for the relationship of the configurations of starting materials and products, which is expressed in modern form in Scheme 2.[75] Astonishingly, few exceptions to the rule are known, and these exceptions can usually be explained as caused by peculiarities in the alcohol structure.

The rule covered the 37 cases of the McKenzie reaction then known and explained the steric course of many analogous cases. To support the postulate that the space demand of the atoms or hydrocarbon residues at the chiral center of the alcohols (S [small] < M [medium] < L [large]) in the neighborhood of the reacting carbonyl is an important factor in determining the steric course of the reaction,

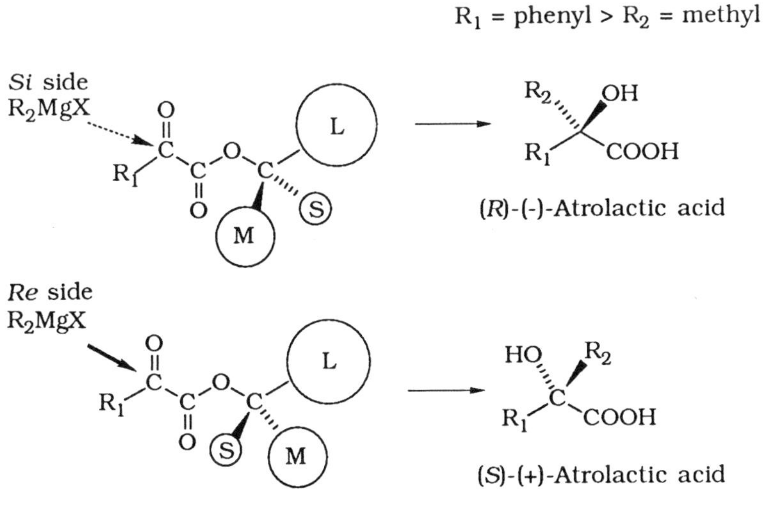

Scheme 2. α-Ketocarboxylic esters of (R)-alcohols react preferentially from the Si face of the carbonyl; for example, if in the ketocarboxylic acid R_1 = phenyl > R_2 = methyl and if the inducing alcohol in which L = 2,4,6-tricyclohexylphenyl > M = methyl > S = hydrogen has the R configuration, the reaction with R_2MgX gives 66% enantiomeric excess of (R)-(−)-atrolactic acid. See reference 76.

we synthesized a series of enantiomeric alcohols with ever larger L residues and showed that we could obtain from their phenylglyoxylic esters and methyl magnesium bromide a substantial enantiomeric excess of atrolactic acid.[76] If the differences S < M < L are significant, the rule permits the derivation, from the configuration of the enantiomer of the α-hydroxycarboxylic acid in excess, of the configuration of the inducing alcohol and vice versa. We used the rule to establish the unknown configuration of several natural secondary alcohols via the atrolactic acid synthesis.[77] Finally, we determined also the configuration of an α,β-hydroxycarboxylic acid formed by oxidative degradation of linalool, a tertiary alcohol.[78] Through a trivial but unfortunate error, we derived the wrong configuration, although the rule was valid and the experimental results were correct. The error was uncovered by John Cornforth, and I corrected it in my only joint publication with Cornforth and his wife.[79]

Among asymmetric reactions, we also investigated the asymmetric cyanohydrin synthesis catalyzed by enantiomeric bases. We showed that catalysis by chiral bases (which, in accordance with the classic concept of Arthur Lapworth, produce the necessary cyanide ions) and the stereoselective course of the reaction can be separated by using an achiral tertiary amine as catalyst while generating the stereoselectivity with a chiral (e.g., quaternary) ammonium ion. For instance, cinnamaldehyde and hydrogen cyanide form an optically active cyanohydrin even with an achiral base such as triethylamine if a chiral quaternary ammonium salt such as quinine bismethiodide is present. The ammonium salt must therefore be a significant component of the transition state or of the intermediate in this stereospecific reaction.[80] This work, in which Max Wilhelm handled large quantities of anhydrous hydrocyanic acid without mishap, is unfortunately given too little attention in books on reaction mechanisms.

Stereoselectivity of Microbial and Enzymic Reactions

Stereoselectivity of asymmetric syntheses is only a step away from that of microbial and enzymatic reactions, a step I took in the mid-1950s. First, we investigated the stereoselective reduction of carbonyl groups by using cultures of microorganism. Augustin Prieto, who joined us from J. C. Gunsalus's group (Urbana, Illinois), brought

Rita and Sir John ("Kappa") Cornforth.

the know-how, and Werner Acklin, Ruzicka's stepson, not only participated but also supervised others who carried out microbial reactions. The microbial cultures were supplied by CIBA, which was then testing many microorganisms for their ability to introduce hydroxyl groups in steroids by oxidation in order to obtain a starting material for corticoid hormones. We used mainly mono- and bicyclic ketones and diketones with carbonyls in six-membered rings as substrates.

In preliminary studies, we first tested the microorganisms for stereoselectivity and then studied in more detail those microorganisms that reduced the various substrates with high enantioselectivity. For specification of stereoselectivity, the stereoisomeric decalones were particularly suitable as substrates. As an example, the reduction products from the stereoisomeric 1,4-decalindiones with *Curvularia falcata* cultures are shown in Scheme 3. Remarkably, the ten stereogenic carbon atoms formed by microbial reduction all possess the same S configuration, independent of the configuration of the other stereogenic centers in the molecule or of whether the hydroxyls are in the axial or the equatorial position. Hydrogen always approaches carbonyl from the *Re* side. Analogous results were obtained with 1- and 2-decalones.

Re side   S configuration

The carbon skeletons of all the reduction products can be fitted into a diamond lattice in which the position of the carbon carrying oxygen is the same. In this way, a diamond lattice section that specifies the stereoselectivity of the particular microorganism is obtained. If one assumes that this section represents the space in which substrates and products can be accommodated in the enzymes of the microorganism, one can then predict, on the basis of the diamond lattice section, what other substrates can be reduced by the microorganism and its enzymes and which products would be formed.

Our approach to the stereoselectivity problem of microbial and enzymatic reactions was at first purely structural. We drew our conclusions from the configurations of substrates and the reduction products obtained from them.[81] However, in the late 1950s, it became clear that we had to deepen and strengthen our conceptions by isolating the enzymes and carrying out kinetics studies. Our first achievement was to show that *Curvularia falcata* contained at least two $NADPH^+$-dependent oxidoreductases with the same Re stereospecificity. The major component of the mixture favors the transfer of hydrogen into the axial position (a–Re–enzyme), and the minor component favors the equatorial position (e–Re–enzyme).

Isolation of sufficient quantities of oxido-reductases from microbial cultures was a laborious and time-consuming task. When Minor J. Coon came to Zurich to spend his sabbatical leave with us in 1961, we suggested to him the isolation and purification of such an enzyme. He showed that decalones could be reduced with pig liver homogenate as readily as with microbial cultures. Hans Dutler, who joined the enzyme group and ultimately became its leader, later isolated the enzyme and identified it as a fatty-acid synthetase complex.[82] On the basis of the stereochemistry of the products from decalones, we recognized the oxido-reductase component of this

PRELOG *My 132 Semesters of Chemistry Studies* 67

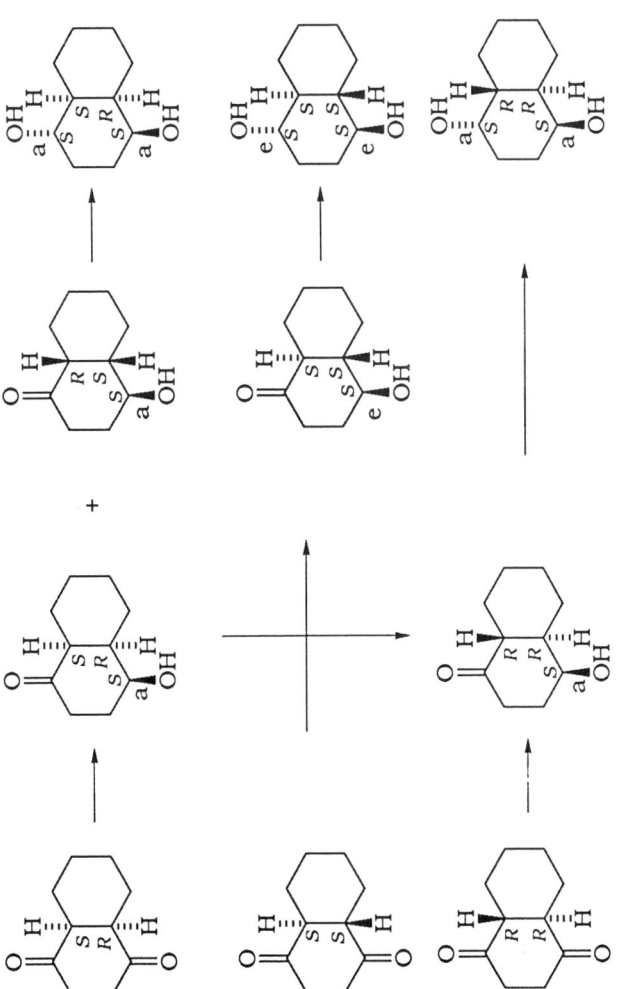

Scheme 3. *Reduction of stereoisomeric 1.4-decalindiones with Curvularia falcata.*

enzyme as an e–Re–enzyme that favors transfer to the equatorial position on the Re side of the carbonyl. Finally in 1973, one of the oxido-reductases isolated from cultures of *Mucor javanicus* was found to be an e–Si–enzyme.[83] Work with isolated enzymes and particularly the kinetics studies occurred when I did not have the material means, nor the number of co-workers, nor the necessary energy to pursue it with enough intensity. I therefore gave up this very promising field as I had given up many others before.

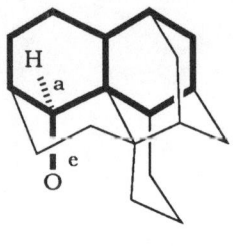

a-Re-enzyme

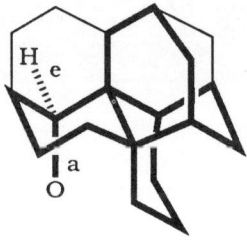

e-Re-enzyme

Laboratory Head (1957), Collegial Chairmanship (1964), and Retirement (1976)

At this point, I want to describe some of the changes in our laboratories and my personal fate between 1957, when I became Ruzicka's successor and laboratory director, and 1976, when I retired. Concurrent with my appointment in 1957, my two younger colleagues and *Privatdozenten* Emil Hardegger and Oskar Jeger were appointed associate professors. On the day Ruzicka stepped down, a third associate professor, Jack D. Dunitz, joined us from the Royal Institution in London.

Soon it became possible to promote other *Dozenten*, Edgar Heilbronner, Albert Eschenmoser, Duilio Arigoni, and Wilhelm Simon, first to associate professors and later, all of them, one by one, to full professors. We divided the teaching duties among us, and Bruno Engel competently and loyally carried out the daily administration of the laboratory. My major task of procuring the financial

Discussing stereospecificity with Duilio Arigoni and Frank Westheimer at the 1976 Bürgenstock Conference.

means for research was not difficult, because 1957 was the year of Sputnik and the West began to support scientific and technical research far more strongly than before.

Both for our laboratory and for me personally, it was a heavy loss when our administrator Bruno Engel for unknown reasons took his own life in the spring of 1964. I had to take over some of his duties and found that I did not enjoy administration. It is not only time consuming and thus interferes with research but also makes it necessary to make decisions about the fate of individuals whom one does not know sufficiently. That type of decision making needs a desire for power, which, my colleagues agree, I totally lack.

On my suggestion, the president of the ETH in 1965 established a collegial leadership for our laboratory in which all appointed professors participate, an arrangement now common but at that time unusual and not accepted by most other laboratory heads at the ETH. This organizational change contributed greatly to keeping my colleagues at the ETH, despite very tempting offers from other universities. Only Heilbronner left us to become professor of physical chemistry at the University of Basel. However, I do not want to give

Oskar Jeger, Jack Dunitz, and Albert Eschenmoser at the symposium at the Weizmann Institute in Rehovot, Israel, in 1986 organized on the occasion of Vladimir Prelog's 80th birthday. (Reprinted with permission from reference 110. Copyright 1986 Chimia.)

the impression that mine was the only influence modifying the hierarchic structures of the ETH. The physicists were striving for similar changes in institutional leadership at the same time. A system changes when the time is ripe. In earlier days, it was the custom at ETH to have one full professor (*Ordinarius*) at an institute who also headed the institute. Next to the laboratory head, Ruzicka, I was, as far as I know, the first so-called *personal Ordinarius*. After 1960, several other institutes had a number of full professors.

My freedom from administrative duties allowed me to devote myself more fully to research and to be more active in other fields. In 1960, I was elected to the governing board of CIBA, and when CIBA merged with J. R. Geigy to become the largest Swiss chemical concern, I continued as a member of the board of Ciba–Geigy until 1978. In this way, I gained a deeper insight into the functioning of an enterprise that had supported me generously from the time I arrived in Switzerland and that I had known for many years from another perspective, as a scientific consultant.

With Kamila Prelog, looking at the A. W. Hofmann Medal in Berlin in 1967. (Reprinted with permission from reference 113. Copyright 1986 Chimia.)

Those of my younger colleagues who became my successors continue with the tradition of the laboratory and contribute significantly to the progress of our science. Their achievements and their friendship have helped greatly to make the semesters after my retirement as pleasant as possible.

On the board, of which I was a member for 18 years, sat many impressive personalities from public life whom I would hardly have come to know as a pure scientist. By name I want to mention only Carl J. Burckhardt, with whom I often traveled by train to Zurich after the meetings in Basel. I will never forget the conversations with this great historian, writer, and statesman. Burckhardt had been United Nations Commissioner in Danzig and vice-president of the

With chemistry students, meeting of Nobel laureates in Lindau in 1977.

Red Cross during the Second World War, when he had to negotiate with leading Nazis. On the other hand, it followed from our talks that his knowledge of physics was pre-Galilean, echoing C. P. Snow's *Two Cultures*.

As a board member of CIBA, I could actively support the founding of the Woodward Institute in Basel. Ruzicka very early had played with the idea of bringing R. B. Woodward to the ETH. After Woodward had gained the much-coveted position as a research professor at Harvard, we gave up hope of attracting him to Switzerland, where no pure research professorships existed. Finally, the director of CIBA, Albert Wettstein, proposed the founding of an institute within the framework of the Basel research laboratories, where Woodward could conduct research on topics of his choice. The institute was to be mainly an advanced school of synthetic chemistry, but any practical, useful results would accrue to the company. Woodward agreed to this plan, which came to fruition in 1960. For me, it was a welcome opportunity to meet him often in Basel, especially after he too was elected to CIBA's board.

In 1963, CIBA opened an additional research institute in Goregaon, north of Bombay in India, and several scientists were

Magic with Koji Nakanishi at the 14th IUPAC Natural Products Symposium in Poznan, Poland, in 1984. (Photograph courtesy of Karel Ubick.)

invited for the opening celebration, among them Alexander Todd and Woodward. After the opening, we traveled together for two weeks in India and were deeply impressed with the past and present of this country. We were Nehru's guests in New Delhi, lectured in Bombay and Poona, and met many Indian chemists.

During the period I have been describing, I traveled quite extensively in Europe, North America, Asia, South Africa, and Australia, mainly in connection with congresses, symposia, lecture tours, and guest professorships. Some of the lecture tours and guest professorships lasted several weeks, such as the Firth lectures in Sheffield, the Todd lectures in Cambridge, the Johnson lectures at Yale, the Baker lectures at Cornell, the Folkers lectures in Urbana and Madison, the Andrews lectures in Sydney, the Pacific Coast lectures in the western United States, the Sigma Xi lectures in the eastern United States, and the guest professorship in Haifa. A short time ago, I estimated that I have lectured at more than 150 places, in some of them several times. This is even more astonishing considering that

With Ernest L. Eliel at the Nomenclature Conference in London organized by the CIBA Foundation in 1968.

I do not even speak my mother tongue without an accent (of unknown origin) and that I do not speak any language flawlessly. Since 1959, I have been a citizen of Zurich and thus also of Switzerland. Meanwhile, through many trips and contacts abroad, I have become something of a world citizen. I am especially grateful to my adopted country for its generosity in allowing me to be one.

The CIP System

After the war, for many years as successor to my doctoral mentor, Votoček, I was a member of the IUPAC commission that sought to standardize and rationalize the nomenclature of organic chemistry. The commission was chaired and carefully guided by Peter Verkade. We met either in Scheveningen near Verkade's residence in The Hague or at places such as New York or Stockholm where IUPAC congresses were taking place. Most of the members of the commission were editors of chemical literature, such as Robert S. Cahn, the journal editor of the London Chemical Society, or Friedrich Richter, the editor of *Beilstein's Handbuch*. By working with the commission, I became sensitized to nomenclature problems, and the acquaintance with Cahn and Richter was later of great significance in the conceptualization and acceptance of the CIP (Cahn–Ingold–Prelog) system, which is now generally used for the unambiguous specification of stereoisomers.

In my work with natural products, as well as with the steric course of chemical and biochemical reactions in which stereoisomeric substrates, transition states, and products play an important role, the lack of a general unambiguous system for specifying and identifying stereoisomers was becoming ever more noticeable.

From the time that Emil Fischer determined the relative configurations of the asymmetric carbon atoms of important natural products, sugars, and α-amino acids through their chemical relationships and specified them by the descriptors *d* and *l* (later replaced by D and L), chemists have been battling for a unique descriptor system for stereoisomers, with the result that C. Buchanan wrote in 1951:[84]

> Although the relative configurations of many optically active compounds could be established, it is impossible to allocate the compounds to D- or L-series without ambiguity. This is true even in carbohydrate chemistry. . . . Surely it would be better to abandon the use of prefixes D and L, except for those parts of the carbohydrate and α-amino acids fields in which they do serve a useful purpose.

At about the same time, Robert Cahn and Christopher Ingold suggested a new procedure for assigning the descriptors D and L to asymmetric carbon atoms.[85] At the 1954 annual meeting of the Chemical Society in Manchester, a symposium on "Dynamic Stereochemistry" was held in which I actively participated. At the end of the meeting on April 2, ICI (Imperial Chemical Industries) invited the participants to a reception and dance in their Hexagon House in Blackley. Among the few nondancers were Ingold and Cahn, as well as my humble self. We discussed the new suggestion to use atomic numbers and topological distances from asymmetric atoms to determine the sequence of their ligands. As I strongly criticized some

Robert Sidney Cahn (1899–1981) at Bürgenstock in 1966.

of their proposals, the British colleagues invited me to join them to work out better ones. I immediately accepted their invitation, without giving much thought to the far-reaching consequences.

Through correspondence and many meetings in England and Switzerland, we came to agreement about the principles on which the new system should be built. Two years later, in 1956, we were ready to present the system to the profession in the Swiss journal *Experientia*.[86] The system was witheringly criticized from various sides, among others by Sir Robert Robinson, who wrote: "Though it may be a 'voice crying in the wilderness' it seems desirable to point out that the symbols R and S, recently proposed in order to denote absolute configurations around a single asymmetric carbon atom, are unnecessary additions to nomenclature."[87] I believe that this criticism derives from Robinson's long-standing feud with Ingold.

Sir Christopher Ingold (1893–1970) at Bürgenstock in 1966.

It was of the utmost importance that Friedrich Richter and his co-worker Otto Weissbach adopted the system for *Beilstein's Handbuch* and applied it to countless stereoisomers. In the process, they encountered some inadequacies of the system, which fortunately were not fundamental ones, and shared them with us. Their criticism and that of several other colleagues led us to write a second paper. It was clear to us that our descriptors R and S specified the chirality of parts of the molecule that we called chirality elements: centers, axes, and planes of chirality. Our second paper,[88] which appeared under the title "Specification of Molecular Chirality" in German and English in *Angewandte Chemie*, contributed much to making the concept of handedness or chirality, formulated by Lord Kelvin,[89] so popular among chemists even if sometimes, unfortunately, wrongly applied. In the second paper, we suggested, along with the descriptors R and S, descriptors for chiral conformations M and P, which derive from an earlier paper in *Experientia*, in which William Klyne and I attempted to name the important types of conformations rationally.[90] Both *Experientia* papers belong to the "Citation Classics".

After Ingold and Cahn died, I, with Günter Helmchen, wrote a third lesser known paper on the CIP system in *Angewandte Chemie*, in which the geometric basis of the system is discussed, the modus operandi is more sharply defined, and certain modifications are suggested.[91] The CIP system has proved itself; it can even be taught to computers and has found entry into all comprehensive textbooks of organic chemistry. Moreover, as shown in a paper by Dieter Seebach and myself, the system is useful for describing unambiguously the steric course of "asymmetric" reactions.[92]

For me, this involvement with the specification of stereoisomers was particularly worthwhile in that I was forced to think about the extent and limits of stereochemistry.[93] The collaboration with the two British chemists and the many conversations and discussions with them, which were not confined to technical matters, made me acquainted with the English way of thinking and broadened my horizons.

Although not part of the CIP system, another nomenclature episode might be worth mentioning here. In 1953, Barton, Hassel, Pitzer, and I published a joint paper in which we proposed the terms axial and equatorial (and the symbols a and e) for the two classes of extracyclic bonds of a cyclohexane ring.[94] In fact, as pointed out in the article, Ingold had suggested to us the term "axial". Prior to these suggestions, there had been no commonly accepted nomenclature. It was important that all protagonists in the field agree about these terms. I was a messenger between Hassel and Pitzer, between whom relations were strained. It took some skill to persuade them to sign the paper. That was my only contribution.

Stereochemical Problems

So far I have described in detail how I arrived at my research topics because I believe that this account is instructive in both positive and negative respects. Looking back, I find that my motivations were mainly based on chance, were naive or emotional, and were seldom rational. The stereochemical studies carried out in connection with the formulation of a general system for specifying stereoisomers (the CIP system) were in this respect an exception. The question arose as to whether we know all the types of stereoisomers and whether stereoisomers are possible that have not yet been realized despite the millions of organic compounds known.

Considering this question, which relates to the scope and limits of static stereochemistry, we carried out work that I have not yet mentioned. In 1964, a novel type of stereoisomerism, cyclostereoisomerism based on two-dimensional chirality of rings, was conceived. As examples of cyclostereoisomers, the cycloenantiomeric cyclohexaalanyls 83 and 84 were prepared by Hans Gerlach and Yuri Ovchinnikov,[95] the first Russian guest in our laboratory.

For political reasons, both before and after the Second World War, contacts with Russian chemists were rare. We met them only at a few international congresses. To increase contacts, I arranged at such a gathering with the leading Russian natural products chemist, Mikhail Shemyakin, that his co-worker, Yuri Ovchinnikov, would spend some months with us.

In connection with considerations about *pseudoasymmetry* in chemistry, new terms were coined for pseudoasymmetric stereogenic entities: pseudoasymmetric centers, axes, and planes. Because stereoisomers with pseudoasymmetric axes and planes were not known, Günter Helmchen synthesized compounds 85–88 as examples.[96] In later research I extended the concept of *geometric enantiomerism*, coined by R. E. Lyle and G. G. Lyle for certain unsaturated stereoisomers, to analogous cyclic compounds, and we produced examples of this type of stereoisomerism, such as the enantiomers 89 and 90.[97] The rotation-symmetric chiral vespirenes 91, showing D_2 symmetry and derived from 9,9-spirobifluorene, were made specifically for chiroptical investigations carried out by Günther Snatzke.[98]

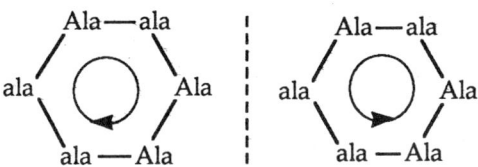

83   84

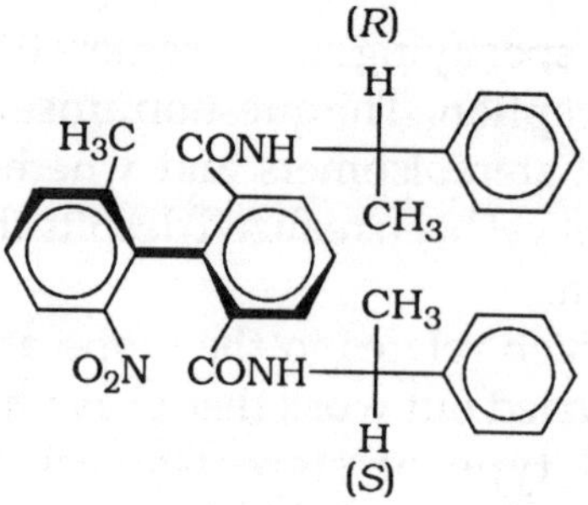

85 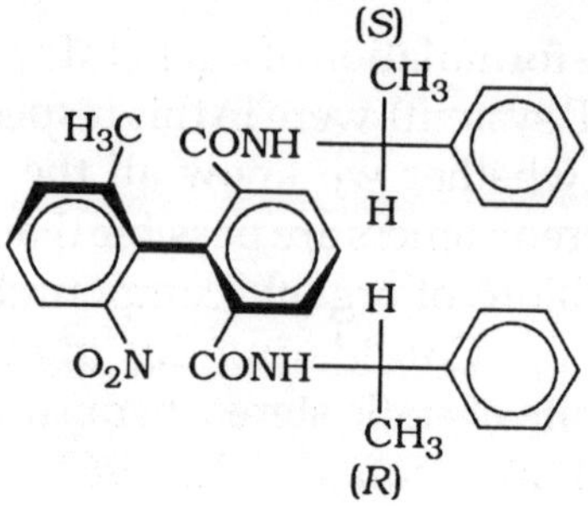 **86**

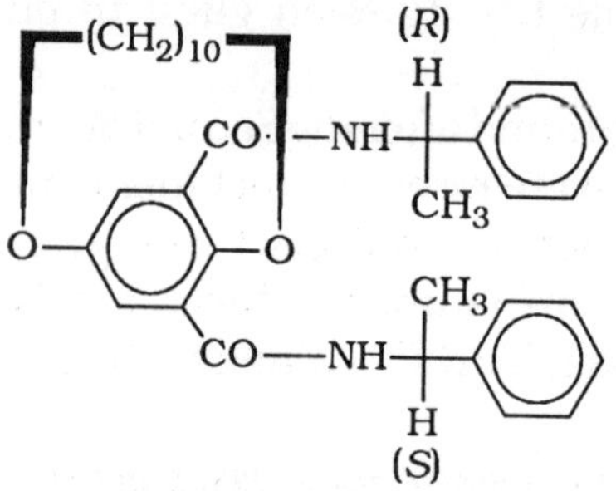

87 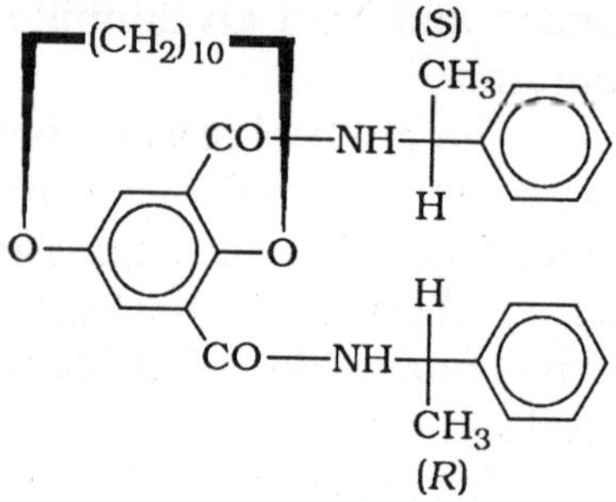 **88**

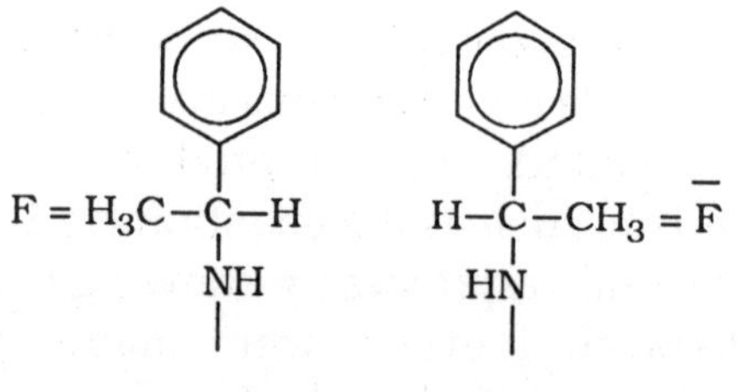

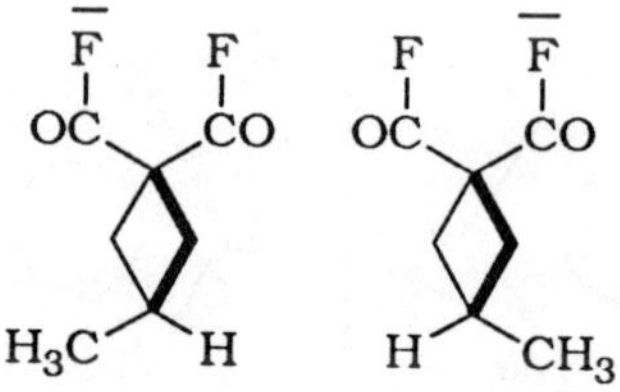

$F = H_3C-\overset{\text{Ph}}{\underset{\underset{|}{NH}}{C}}-H$ $\quad H-\overset{\text{Ph}}{\underset{\underset{|}{HN}}{C}}-CH_3 = \overline{F}$

89 **90**

Yuri Ovchinnikov (1934–1988) at Bürgenstock in 1966.

Since my retirement, I have been registered as a postdoctoral student at ETH and have now (September 1990) completed the 132nd semester of my chemical studies. I have confined myself to a very narrow research field in which a younger postdoctoral student (usually from Zagreb) does the experimental work. It has to do with the separation of enantiomers by distribution between two liquid phases. Wilhelm Simon, who, in our laboratory, carried out fundamental studies on ion-specific electrodes, also constructed the first electrodes that are enantioselective for chiral ammonium ions.[99]

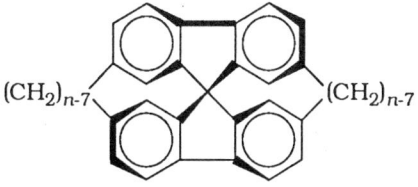

n = 13, 14, 15

Using the experience gained when we synthesized the chiral vespirenes **91**, we synthesized the chiral crown ethers **92** and **93**, which contain 9,9-spirobifluorene as the chiral building block.[100] These compounds were then incorporated in polyvinyl chloride membranes to construct enantioselective electrodes. Their enantioselectivity for chiral ammonium salts was only moderate; in contrast, a significant enantioselectivity was shown in some cases by the bis(9,9-spirobifluorene) crown ethers **94a** and **b**.[101] As byproducts in the preparation of the latter compounds, we obtained the esthetically appealing oligomeric crowns **95a** and **b** with 78- and 104-membered rings, respectively.[102]

When we prepared tartaric esters of higher alcohols as potential chiral plasticizers for polyvinyl chloride electrode membranes, we discovered that these easily available compounds were also

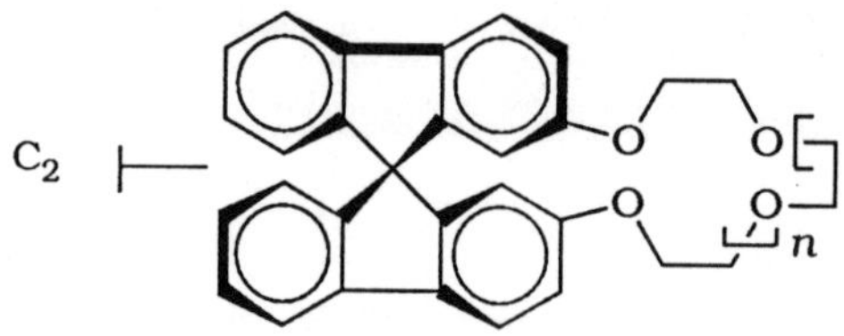

$n = 1, 2, 3$

92

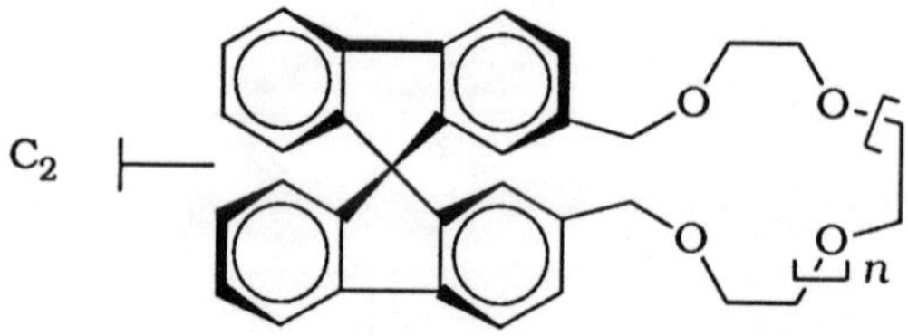

$n = 1, 2, 3$

93

94a, $n = 1$

94b, $n = 2$

95a, $m = 1$

95b, $m = 3$

Wilhelm Simon in Zurich.

stereoselective ionophores for α-aminoalcohol and α-aminoacid ester salts. With these esters as components of lipophilic phases, it is possible to separate enantiomers by partition between aqueous and lipophilic phases. We are now focusing on the mechanism of this enantioselectivity and its applications.[103]

Leopold Ruzicka, 88, congratulating Vladimir Prelog on his receipt of the Nobel Prize, Zurich, October 1975. (Reprinted with permission from reference 112. Copyright 1986 Chimia.)

In Retrospect

As an added occupation, since 1977, I have been organizing Ruzicka's literary estate and, together with Oskar Jeger, have written his biography for the *Biographical Memoirs of the Royal Society* and a shorter version for *Helvetica Chimica Acta*.[104]

In writing the essay about Ruzicka, I was stimulated to think about my own past. I discovered, for instance, that I have had about 100 doctoral students and a similar number of other co-workers and that I have published some 400 papers. Looking back, I became rather clearly aware that all in all for someone born in Sarajevo, I have been very lucky and I cannot imagine alternate circumstances in my life that would have allowed me to achieve more.

At Bürgenstock in 1989. (Photograph courtesy K. Zimmerman.)

Every autobiography is something of an *apologia pro vita sua*. I cannot justify my activities in a better way than by quoting the words which Sir John Cornforth used on behalf of both of us when he expressed our thanks on being honored with the 1975 Nobel Prize in chemistry:

> Our backgrounds, and the experience that has shaped us as scientists, are very different. We were born, and we grew up, on opposite sides of the globe. What we have in common is a lifelong curiosity about the shapes, and changes in shape, of entities that we shall never see; and a lifelong conviction that this curiosity will lead us closer to the truth of chemical processes, including the processes of life.[105]

EX LIBRIS

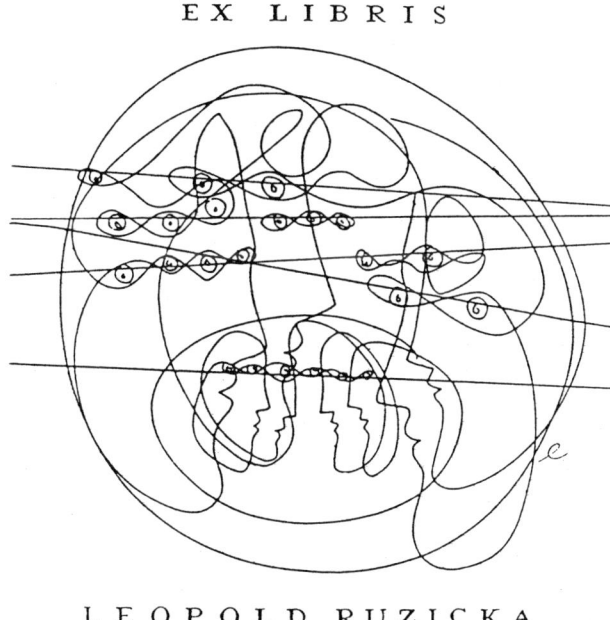

LEOPOLD RUZICKA

Ruzicka's ex libris by Hans Erni.

Epilogue

The Ex Libris of Hans Erni

On the occasion of Leopold Ruzicka's 65th birthday, his laboratory wanted to give him a present. Our administrator at the time, Bruno Engel, had the good idea of asking the Swiss artist and printmaker Hans Erni (b. 1909) to make an ex libris for Ruzicka's sizable library. Ruzicka's phrases "Many eyes see more than two" and "No professor is better than the sum of his co-workers" served as text.

When I received the 1975 Nobel Prize, many friends, colleagues, and acquaintances wrote me, and I wanted to thank them by sending them a small engraving, which I asked Erni to create. I talked with him about chirality and told him that for dealing with it in chemistry human intelligence is needed plus the left and right hands and the two simplest chiral objects of three-dimensional space, the enantiomorphic tetrahedra. Erni thereupon drew several dozen

Prelog's ex libris by Hans Erni.

sketches representing these attributes. One of them was executed as an engraving and later also became my ex libris. The girl in the design acquired the name *Rassi-maid* (racemate) and appears also in Erni's lithograph for R. B. Woodward's 60th birthday.

A Triptych

Prelog, Robert Woodward, and Carl Djerassi in the Baltic Sea at the 1970 IUPAC Symposium on Natural Products in Riga, USSR.

To Prelog, these photographs showed that "we are the solid fundaments of His glory." According to Djerassi, the chemists were "staging the supposed superiority of Harvard over ETH and Stanford." (Photographs courtesy D. E. Koshland, Jr.)

I have a reputation as a storyteller. I have not told any stories in my text until now, because stories are mostly apocryphal, or at least, they tend to deviate from the truth, and I wished to avoid that. Notwithstanding, at the close I want to tell four anecdotes that are not so commonly known about four great organic chemists.

Emil Fischer

In 1940, when Germany was already at war, the German research leader and industrial chemist Fritz Hofmann visited me in Zagreb. He was well known to me for his technical synthesis of isoprene and its polymerization to rubber during the First World War. We both hated the war and the political situation, so we spoke about the past, and he told me, among other things, about his encounter with Emil Fischer. Hofmann's superior, Carl Duisberg, the great industrial leader and founder of the I. G. Farben, had once told him that he wanted to tie *Geheimrat* Fischer to their firm. He proposed that Hofmann was to visit Fischer in Berlin, persuade him that Fischer was coinventor of the successful drug bromural (α-bromoisovaleryl urea), and offer him 5% of the sales as royalties. In fact, Fischer's only contribution in this respect was that he had been the first to describe in the literature α-bromoisovaleryl chloride, the starting material for the synthesis. Hofmann, who didn't like this commission, tried to persuade Duisberg that Fischer would never accept such an offer, but Duisberg remained firm. After a sleepless night in a sleeping car on the train to Berlin, during which he contemplated what to say, Hofmann met Fischer next day in his office. The result of his action was unexpected; Fischer demanded 10% royalties instead of 5%. This despite the statement of Emil's father, Lorenz, a shrewd businessman, about him: *"Der Junge ist zum Kaufmann zu dumm, er soll denn in Gottes Namen studieren."* (The boy is too stupid to go into business; so in God's name, let him study.)

Roger Adams

In the spring of 1950, during my first visit to the United States, I visited the Department of Chemistry of the University of Illinois in Urbana, a stronghold of organic chemistry. I was the guest of Roger Adams, then the most impressive and influential American organic chemist. He traveled with me to Chicago in a lounge car of the Panama Express, which was full of lawyers. One of them, who evidently knew Adams from previous trips, asked him loudly, "Roger, you're a chemist, aren't you? For heaven's sake, can you tell me what is a `kae'sh n'?" Adams replied, "If you want to impress the experts in the court you better pronounce it cation (`kad'i n')," and then he added, "My fee is $10." Everybody laughed approvingly, and Adams pocketed the money.

Robert Robinson

As I already mentioned, Robinson disliked the CIP system for the specification of molecular chirality. When we once met at Zurich airport on the way to Israel to celebrate the 25th anniversary of the Weizmann Institute, the first words we exchanged were the following: Robinson: "Hello, Katchalsky. What are you doing here in Zurich?" I: "Excuse me, Sir Robert, I am only Prelog, and I live here." Robinson: "You know, Prelog, your and Ingold's configurational notation is all wrong." I: "Sir Robert, it can't be wrong. It is just a convention. You either accept it or not." Robinson: "Well then, if it is not wrong, it is absolutely unnecessary."

Robert Burns Woodward

Woodward and Sir Robert Robinson held each other in high esteem, but they did not always get along well. One day, probably in the late 1950s, when Woodward visited us in Zurich he said to me, "You know, Sir Robert is a bad old man." I: "How can you say something like that about the greatest living organic chemist?" Woodward:

"Yesterday, I spent the whole day with him in Oxford. He doesn't communicate much with me any more about chemistry, but he does talk about individuals. He didn't find a single good word for any chemist in the whole world." I: "Perhaps he was in a bad mood; that doesn't mean that he is evil." Woodward, irritated by my contradicting, replied, "Indeed, I didn't tell the truth. He said something nice about you." I, suspecting something wrong, asked, "What did he say?" And Woodward triumphantly replied, "Prelog is not a good chemist, but he is a nice person." How true!

Editor's note: As this volume was advancing toward the last stages of production, I learned from Professor Prelog that he had written a number of additional anecdotes at the request of Albert Eschenmoser. Prelog kindly sent nine of these stories to me without any intention of incorporating them into this volume. O. T. Benfey translated a number of them into English, Prelog agreed that three of them could be added to his volume, and ACS Books generously allowed them to be included at the last moment.

Ruzicka in Paris

Ruzicka was drawn to beautiful women. Long ago, he told me the following story. I don't remember all the details, but by and large it is accurate.

When, in 1927, Richard Kuhn succeeded Hermann Staudinger in the ETH chair for general and analytical chemistry, Ruzicka saw no future for himself at ETH. He therefore accepted a position with the Geneva perfume manufacturers M. Naef & Co., formerly Chuit, Naef & Co. While there, he was invited by the Société Chimique de France in Paris to give a lecture about his recently discovered large-ring compounds with musk odor.[106] Ruzicka's French was not very good, and he had never before been in Paris. His friends in Geneva were worried. They translated his German manuscript, and he memorized it. They assigned the firm's Paris representative, Monsieur Chevron, to look after Ruzicka. Chevron and his young, pretty wife did their best to make Ruzicka's stay pleasant; they showed him

Paris and were also present at his lecture. The presence of the beautiful woman was a stimulus to Ruzicka and helped him master his linguistic difficulties. He was sure he had made a deep and lasting impression on her.

When, about 10 years later, Ruzicka gave another lecture in Paris, this time at the Congrès du Palais de Découvert, on the architecture of the polyterpenes,[107] the Chevrons again looked after him. They brought him to the lecture hall, and Ruzicka said to the still very pretty Mme. Chevron, "Of course you are staying for my lecture?" Embarrassed, she answered, "You know, Professor Ruzicka, your French has improved so much that it wouldn't be fun any more." Ruzicka told me it was a great disappointment for him.

A Glutamic Acid Synthesis

The following story was told to me by my mentor, Rudolf Lukeš. Emil Votoček, under whom both of us obtained our doctorates, had begged for several grams of "succinic acid hemialdehyde" from Carl ("Ozone") Harries to use it to make glutamic acid by the Strecker method. A doctoral candidate completed the synthesis in unimpressive yield and handed the product to Votoček, who himself determined the melting point. Because it was suspiciously high, considering that the enantiomeric glutamic acids melt higher than the racemic mixture, he also determined the optical rotation. This showed that he had the L-enantiomer rather than the racemic acid. He accused the graduate student of having messed up the difficult-to-obtain hemialdehyde and handing him the natural glutamic acid instead. The student swore that it was the product of his synthesis. Votoček thereupon wrote William J. Pope of Cambridge, the "pope" of stereochemistry, who assured him that a spontaneous resolution was out of the question. Votoček then gathered all the members of his laboratory and before the assembled company, he drove the graduate student, who swore that he was innocent, forever out of the lab. I don't know this person's name or what became of him. Very likely, however, he suffered a great injustice, for today we know that racemic glutamic acid easily yields enantiomers by spontaneous resolution.

The Rifamycin Story

In 1960, at the suggestion of Albert Wettstein, I had a visit from the research director of Lepetit S.A. (Milan), Piero Sensi, who proposed that I collaborate with his research group on the structure of the effective new antibiotics discovered in Milan, rifomycins. He could supply us with sizable amounts for our use. Needless to say, I immediately agreed to this tempting offer that included financial support.

At the same time, a young Austrian, Wolfgang Oppolzer, registered as one of my new doctoral students. He had been warmly recommended by his professor in Vienna, Fritz Wessely.

The two developments are a confirmation of Leopold Ruzicka's assertion that a researcher doesn't look for problems to tackle, but that problems find those who will work on them. Oppolzer, in a race with X-ray crystallographers, solved his problem brilliantly.

Sensi, who was intensely interested in the structure of rifomycins, had, it turned out, also asked the Yale X-ray crystallographer Alexander Tulinsky to tackle the structure problem. Time was therefore of the essence. Tulinsky, however, had bad luck with the crystals he examined. They always disintegrated shortly after X-ray exposure. He finally gave up the problem, and our work became more relaxed.

Soon, however, Sensi found a second X-ray crystallographer, Alessandro Vaciago, to work on rifomycins in Rome. In the meantime, we had completed and published our work.[108] I even boldly asserted in a public lecture[109] that we were confidently awaiting the X-ray crystallography verdict.

One day it came. Vaciago telephoned me that he and Mario Brufani had determined the structure of the p-iodoanilide of rifomycin B and thus also of other related rifomycins. To my question whether his and our structures agreed, he responded positively, to my great relief. The photograph of the model of his structure was already on its way to Zurich. A detailed inspection of the photo, however, turned out to be disappointing. The rifomycins, now rebaptized rifamycins for patent law reasons,[114] contain an $\alpha,\beta,\gamma,\delta$-doubly unsaturated carboxyamide group. On the basis of NMR spectra, we decided that the α,β-double bond was cis and the γ,δ bond was trans. In the photograph of Vaciago's model, on the other hand, the first bond was trans and the second cis. Because I could find no reason for this discrepancy, I called Vaciago the next day and asked him if the configurations in his model were absolutely certain. Conscientious checking in Rome then revealed that the model had been taken in a taxi to the photographer, but that in the unpredictable Roman traffic, the model came apart at a sudden halt. In the photographer's studio, the model was put back together—incorrectly. Some chemists feel a sense of inferiority toward X-ray crystallographers and often believe them implicitly, so we were proud of our confidence in our own findings.

References

References

1. Prelog, V. *Chem. Ztg.* **1921**, *45*, 736.
2. Ostwald, W. *J. Chem. Soc.* **1904**, *85*, 506.
3. Lukeš, R.; Prelog, V. *Collect. Czech. Chem. Commun.* **1929**, *1*, 334.
4. Votoček, E.; Prelog, V. *Collect. Czech. Chem. Commun.* **1929**, *1*, 55.
5. Prelog, V.; Dřiža, J. G. *Collect. Czech. Chem. Commun.* **1931**, *3*, 578. Ibid. **1932**, *4*, 32. Ibid. **1933**, *5*, 497.
6. Prelog, V.; Kohlbach, D. *Collect. Czech. Chem. Commun.* **1936**, *8*, 377.
7. Ruzicka, L.; Prelog, V. *Helv. Chim. Acta* **1937**, *20*, 1570.
8. Prelog, V.; Stern, P.; Seiwerth, R.; Heimbach-Juhász, S. *Naturwissenschaften* **1940**, *28*, 750.
9. Prelog, V.; Seiwerth, R.; Heimbach-Juhász, S.; Stern, P. *Chem. Ber.* **1941**, *74*, 647. Rabe, P.; Hagen, G. *Chem. Ber.* **1941**, *74*, 636.
10. Proštenik, M.; Prelog, V. *Helv. Chim. Acta* **1943**, *26*, 1965.
11. Prelog, V.; Kohlbach, D.; Cerkovnikov, E.; Režek, A.; Piantanida, M. *Liebigs Ann. Chem.* **1937**, *532*, 69.
12. Prelog, V.; Cerkovnikov, E. *Liebigs Ann. Chem.* **1936**, *525*, 292; **1937**, *532*, 83. Prelog, V.; Cerkovnikov, E.; Ustricev, G. *Liebigs Ann. Chem.* **1938**, *535*, 37. Prelog, V.; Heimbach, S.; Cerkovnikov, E. *J.*

Chem. Soc. **1939**, 677. Prelog, V; Heimbach, S. Chem. Ber. **1939**, 72, 1101. Prelog, V. et al. Liebigs Ann. Chem. **1940**, 545, 229, 231, 243, 247, 256, 259.

13. Prelog, V.; Božičević, K. Chem. Ber. **1939**, 72, 1103.
14. Prelog, V.; Seiwerth, R. Chem. Ber. **1939**, 72, 1638.
15. Prelog, V.; Komzak, A. Chem. Ber. **1941**, 74, 1705.
16. Prelog, V.; Komzak, A.; Moor, E. Helv. Chim. Acta **1942**, 25, 1964. Prelog, V.; Moor, E. Helv. Chim. Acta **1943**, 26, 846.
17. Prelog, V.; Heimbach-Juhász, S. Chem. Ber. **1941**, 74, 1702.
18. Landa, S.; Macháček, V. Collect. Czech. Chem. Commun. **1933**, 5, 1.
19. Prelog, V.; Seiwerth, R. Chem. Ber. **1941**, 74, 1644, 1769.
20. Ruzicka, L.; Prelog, V. Helv. Chim. Acta **1943**, 26, 975. Prelog, V.; Tagmann, E.; Lieberman, S.; Ruzicka, L. Helv. Chim. Acta **1947**, 30, 1080.
21. Prelog, V.; Ruzicka, L. Helv. Chim. Acta **1944**, 27, 61.
22. Prelog, V.; Ruzicka, L.; Wieland, P. Helv. Chim. Acta **1944**, 27, 66. Prelog, V.; Ruzicka, L.; Meister, P.; Wieland, P. Helv. Chim. Acta **1945**, 28, 618.
23. Claus, R.; Hoppen, H. D.; Karg, H. Experientia **1981**, 37, 1178.
24. Haines, W. J.; Johnson, R. H.; Goodwin, M. P.; Kuizenga, M. H. J. Biol. Chem. **1948**, 74, 925.
25. Prelog, V.; Ruzicka, L.; Steinmann, F. Helv. Chim. Acta **1944**, 27, 674. Prelog, V.; Beyerman, H. C. Helv. Chim. Acta **1945**, 28, 350.
26. Prelog, V.; Führer, J.; Hagenbach, R.; Frick, H. Helv. Chim. Acta **1947**, 30, 113. Prelog, V.; Führer, J.; Hagenbach, R.; Schneider, R. Helv. Chim. Acta **1948**, 31, 1799. Lederer, E.; Prelog, V.; Schneider, R. Helv. Chim. Acta **1948**, 31, 2133. Prelog, V.; Schneider, R. Helv. Chim. Acta **1949**, 32, 1632. Prelog, V.; Vaterlaus, B. Helv. Chim. Acta **1949**, 32, 2082. Ibid. **1950**, 33, 1725, 2262. Prelog, V.; Osgan, M. Helv. Chim. Acta **1952**, 35, 981, 986.
27. Prelog, V.; Geyer, U. Helv. Chim. Acta **1946**, 29, 1587.
28. Schinz, H.; Ruzicka, L.; Geyer, U.; Prelog, V. Helv. Chim. Acta **1946**, 29, 1524.
29. Prelog, V.; Szpilfogel, S. Helv. Chim. Acta **1942**, 25, 1306. Ibid. **1944**, 27, 390.

30. Prelog, V.; Szpilfogel, S. *Experientia* **1945**, *1*, 197.
31. Robinson, R. *Experientia* **1946**, *2*, 28.
32. Robinson, R. *Nature (London)* **1947**, *159*, 263.
33. Prelog, V.; Zalán, E. *Helv. Chim. Acta* **1944**, *27*, 535, 545.
34. Prelog, V.; Wieland, P. *Helv. Chim. Acta* **1944**, *27*, 1127.
35. Janot, M.-M.; Goutarel, R.; Prelog, V. *Helv. Chim. Acta* **1951**, *34*, 1207. Goutarel, R.; Janot, M.-M.; Mirza, R.; Prelog, V. *Helv. Chim. Acta* **1953**, *36*, 337. Janot, M.-M.; Goutarel, R.; Le Hir, A.; Tsatsas, G.; Prelog, V. *Helv. Chim. Acta* **1955**, *38*, 1073.
36. Goutarel, R.; Janot, M.-M.; Prelog, V.; Taylor, W. I. *Helv. Chim. Acta* **1950**, *33*, 150.
37. Prelog, V.; Wiesner, K.; Khorana, H. G.; Kenner, G. W. *Helv. Chim. Acta* **1949**, *32*, 453. Carmack, M.; McKusick, B. C.; Prelog, V. *Helv. Chim. Acta* **1959**, *34*, 1601. Kenner, G. W.; Khorana, H. G.; Prelog, V. *Helv. Chim. Acta* **1951**, *34*, 1969. Prelog, V.; McKusick, B. C.; Merchant, J. R.; Julia, S.; Wilhelm, M. *Helv. Chim. Acta* **1956**, *39*, 498.
38. Prelog, V. *Angew. Chem.* **1957**, *69*, 33.
39. Boekelheide, V.; Prelog, V.; *Progress in Organic Chemistry*; Cook, J. W., Ed.; Butterworths: London, 1955, Vol. 3, p 267.
40. Prelog, V.; Norymberski, J.; Jeger, O. *Helv. Chim. Acta* **1946**, *29*, 360. Jeger, O.; Norymberski, Y.; Szpilfogel, S.; Prelog, V. *Helv. Chim. Acta* **1946**, *29*, 684.
41. Prelog, V.; Metzler, O.; Jeger, O. *Helv. Chim. Acta* **1947**, *30*, 675. Prelog, V.; Ruzicka, L.; Metzler, O. *Helv. Chim. Acta* **1947**, *30*, 1883. Ruzicka, L.; Prelog, V.; Battegay, J. *Helv. Chim. Acta* **1948**, *31*, 1296. Prelog, V.; Würsch, J.; Königsbacher, K. *Helv. Chim. Acta* **1951**, *34*, 258.
42. Prelog, V.; Jeger, O. In *The Alkaloids*; Manske, R. F.; Holmes, H. L., Eds.; Academic: New York, 1953, Vol. 3, p 247; 1960, Vol. 7, pp 319, 343, 364.
43. Barton, D. H. R.; Jeger, O.; Prelog, V.; Woodward, R. B. *Experientia* **1954**, *10*, 81.
44. Prelog, V.; Frenkiel, L.; Kobelt, M.; Barman, P. *Helv. Chim. Acta* **1947**, *30*, 1741.
45. Stoll, M.; Hulstkamp, J. *Helv. Chim. Acta* **1947**, *30*, 1815.

46. Review: Prelog, V. In *Perspectives in Organic Chemistry*; Todd, Sir Alexander, Ed.; Interscience: New York, London, 1956; p 96.

47. Prelog, V.; Kobelt, M. *Helv. Chim. Acta* **1949**, *32*, 1187.

48. Prelog, V. *J. Chem. Soc.* **1950**, 420.

49. Prelog, V.; Ruzicka, L.; Barman, P.; Frenkiel, L. *Helv. Chim. Acta* **1948**, *31*, 92.

50. Prelog, V.; Wiesner, K. *Helv. Chim. Acta* **1947**, *30*, 1465.

51. Prelog, V.; Wiesner, K. *Helv. Chim. Acta* **1948**, *31*, 270, 877.

52. Anliker, R.; Dvornik, D.; Gubler, K.; Heusser, H.; Prelog, V. *Helv. Chim. Acta* **1956**, *39*, 1785. Prelog, V.; Gold, A. M.; Talbot, G.; Zamojski, A. *Helv. Chim. Acta* **1962**, *45*, 4.

53. Dominguez, J.; Dunitz, J. D.; Gerlach, H.; Prelog, V. *Helv. Chim. Acta* **1962**, *45*, 129. Beck, J.; Gerlach, H.; Prelog, V.; Voser, W. *Helv. Chim. Acta* **1962**, *45*, 620. Gerlach, H.; Prelog, V. *Liebigs Ann. Chem.* **1963**, *669*, 121.

54. Štefanac, L.; Simon, W. *Chimia (Switzerland)* **1966**, *20*, 436. *Microchem. J.* **1967**, *12*, 125.

55. Keller-Schierlein, W.; Mihailović, M. Lj.; Prelog, V. *Helv. Chim. Acta* **1959**, *42*, 305.

56. Cf. review articles: Prelog, V. *Pure Appl. Chem.* **1963**, *6*, 327. Keller-Schierlein, W.; Prelog, V.; Zähner, H. *Fortschr. Chem. Org. Naturst.* **1964**, *22*, 280.

57. Dunitz, J. D.; Hawley, D. M.; Mikloš, D.; White, D. N. J.; Berlin, Yu.; Marušić, R.; Prelog, V. *Helv. Chim. Acta* **1971**, *54*, 1709.

58. Prelog, V.; Oppolzer, W. *Helv. Chim. Acta* **1973**, *56*, 2279, 2287. Cf. review articles: Prelog, V. *Pure Appl. Chem.* **1963**, *7*, 551. Oppolzer, W.; Prelog, W.; Sensi, P. *Experientia* **1964**, *20*, 336. Leitich, J.; Oppolzer, W.; Prelog, V. *Experientia* **1964**, *20*, 343.

59. Comin, J.; Gonçalves de Lima, O.; Grant, H. N.; Jackman, L. M.; Keller-Schierlein, W.; Prelog, V. *Helv. Chim. Acta* **1963**, *46*, 409.

60. Prelog, V.; Schenker, K. *Helv. Chim. Acta* **1952**, *35*, 2044.

61. Cope, A. C.; Fenton, S. W.; Spencer, C. F. *J. Am. Chem. Soc.* **1952**, *74*, 5884.

62. Prelog, V.; Schenker, K.; Küng, W. *Helv. Chim. Acta* **1953**, *36*, 471. Schenker, K.; Prelog, V. *Helv. Chim. Acta* **1953**, *36*, 896.

63. Prelog, V.; Boarland, V. *Helv. Chim. Acta* **1955**, *38*, 1776.
64. Prelog, V.; Speck, M. *Helv. Chim. Acta* **1955**, *38*, 1786.
65. Cf. review article: Prelog, V.; Traynham, J. G. In *Molecular Rearrangements*; De Mayo, P., Ed.; Wiley–Interscience: New York, London, 1963; Vol. 1, p 593.
66. Heck, R.; Prelog, V. *Helv. Chim. Acta* **1955**, *38*, 1541. Brown, H. C.; Ham, G. *J. Am. Chem. Soc.* **1956**, *78*, 5599. Mislow, K.; Borčić, S.; Prelog, V. *Helv. Chim. Acta* **1957**, *40*, 2477.
67. Cf. review articles in ref. 65 and Prelog, V. *Angew. Chem.* **1958**, *70*, 145.
68. Dunitz, J. D.; Prelog, V. *Angew. Chem.* **1960**, *72*, 896. Prelog, V. *Bull. Soc. Chim. Fr.* **1960**, 1433. Dunitz, J. D. In *Perspectives in Structural Chemistry*; Dunitz, J. D.; Ibers, J. A. Eds.; Wiley: New York, 1969; Vol. 2, pp 1–70.
69. Lifson, S.; Dunitz, J. D. In *Potential Energy Functions in Conformational Analysis*; Rasmussen, K., Ed.; Springer: New York, 1985, p 119.
70. Cordes, C.; Prelog, V.; Troxler, E.; Westen, H. H. *Helv. Chim. Acta* **1968**, *51*, 1663. Prelog, V.; Troxler, E.; Westen, H. H. *Helv. Chim. Acta* **1968**, *51*, 1678.
71. Prelog, V.; Boarland, V.; Polyák, S. *Helv. Chim. Acta* **1955**, *38*, 434. Prelog, V.; Polyák, S. *Helv. Chim. Acta* **1957**, *40*, 816.
72. Prelog, V.; Polyák, S. *Suom. Kemistil. B.* **1958**, *31*, 95.
73. Kaarsemaker, S.; Coops, J. *Recl. Trav. Chim.* **1952**, *71*, 261. vanKamp, H., Thesis, Vrije Universiteit, Amsterdam, Netherlands, 1957.
74. McKenzie, A. *Ergeb. Enzymforsch.*, **1936**, *5*, 49.
75. Prelog, V. *Helv. Chim. Acta* **1953**, *36*, 308. Prelog, V.; Meier, H. L. *Helv. Chim. Acta* **1953**, *36*, 320.
76. Prelog, V.; Ceder, G.; Wilhelm, M. *Helv. Chim. Acta* **1955**, *38*, 303. Prelog, V.; Philbin, E.; Watanabe, E.; Wilhelm, M. *Helv. Chim. Acta* **1956**, *39*, 1086.
77. Dauben, W. G.; Dickel, D. F.; Jeger, O.; Prelog, V. *Helv. Chim. Acta* **1953**, *36*, 325. Prelog, V.; Tsatsas, G. *Helv. Chim. Acta* **1953**, *36*, 1178. Feldman, W. R.; Prelog, V. *Helv. Chim. Acta* **1958**, *41*, 2396.
78. Prelog, V.; Watanabe, E. *Liebigs Ann. Chem.* **1957**, *603*, 1.

79. Cornforth, R. H.; Cornforth, J. W.; Prelog, V. *Liebigs Ann. Chem.* **1960**, *634*, 197.

80. Prelog, V.; Wilhelm, M. *Helv. Chim. Acta* **1954**, *37*, 1634.

81. Prelog, V.; Acklin, W. *Helv. Chim. Acta* **1956**, *39*, 748. Acklin, W.; Prelog, V.; Prieto, A. P. *Helv. Chim. Acta* **1958**, *41*, 1416, and subsequent papers of this series summarized in review articles: Prelog, V. In *Ciba Foundation Study Group*; Churchill: London, 1959, Vol. 2. *Ind. Chim. Belge* **1962**, 1309. In *Mechanismen Enzymatischer Reaktionen*, Springer: Berlin, 1964, p 288. *Pure Appl. Chem.* **1964**, *9*, 119.

82. Dutler, H.; Coon, M. J.; Kull, A.; Vogel, H.; Waldvogel, G.; Prelog, V. *Eur. J. Biochem.* **1971**, *22*, 213.

83. Dutler, H.; van der Baan, J. L.; Hochuli, E.; Kis, Z.; Taylor, K. E.; Prelog, V. *Eur. J. Biochem.* **1977**, *75*, 423.

84. Buchanan, C. *Nature* **1951**, *167*, 689.

85. Cahn, R. S.; Ingold, C. K. *J. Chem. Soc.* **1951**, 612.

86. Cahn, R. S.; Ingold, C. K.; Prelog, V. *Experientia* **1956**, *12*, 81.

87. Robinson, R. In *Selected Constants: Optical Rotatory Power*; Vol. 4. Alkaloids; Pergamon: London, Paris, New York, Los Angeles, Frankfurt, 1959, p I.

88. Cahn, R. S.; Ingold, C. K.; Prelog, V. *Angew. Chem.* **1966**, *78*, 413. *Angew. Chem. Int. Ed. Engl.* **1966**, *5*, 385.

89. Prelog, V. *Science* **1976**, *193*, 17.

90. Klyne, W.; Prelog, V. *Experientia* **1960**, *16*, 521.

91. Prelog, V.; Helmchen, G. *Angew. Chem.* **1982**, *94*, 614. *Angew. Chem. Int. Ed. Engl.* **1982**, *21*, 567.

92. Seebach, D.; Prelog, V. *Angew. Chem.* **1982**, *94*, 696. *Angew. Chem. Int. Ed. Engl.* **1982**, *21*, 654.

93. Prelog, V. In *van't Hoff—Le Bel Centennial*; Ramsay O. B., Ed.; American Chemical Society: Washington, DC, 1975; p 179.

94. Barton, D. H. R.; Hassel, O.; Pitzer, K. S.; Prelog, V. *Nature* **1953**, *172*, 1096. *Science* **1954**, *119*, 49.

95. Prelog, V.; Gerlach, H. *Helv. Chim. Acta* **1964**, *47*, 2288. Gerlach, H.; Ovtschinnikov, J. A.; Prelog, V. *Helv. Chim. Acta* **1964**, *47*, 2294. Gerlach, H.; Haas, G.; Prelog, V. *Helv. Chim. Acta* **1966**, *49*, 603.

96. Prelog, V.; Helmchen, G. *Helv. Chim. Acta* **1972**, *55*, 2581. Helmchen, G.; Prelog, V. *Helv. Chim. Acta* **1972**, *55*, 2599, 2612.

97. Prelog, V.; Thix, J.; Srikrishnan, T. *Helv. Chim. Acta* **1982**, *65*, 2622.

98. Haas, G.; Prelog, V. *Helv. Chim. Acta* **1969**, *52*, 1202. Haas, G.; Hulbert, P. B.; Klyne, W.; Prelog, V.; Snatzke, G. *Helv. Chim. Acta* **1971**, *54*, 491.

99. Thoma, A. P.; Cimerman, Z.; Fiedler, U.; Bedekovič, D.; Guggi, M.; Jordan, P.; May, K.; Pretsch, E.; Prelog, V.; Simon, W. *Chimia* **1975**, *29*, 344.

100. Prelog, V.; Bedekovič, D. *Helv. Chim. Acta* **1979**, *62*, 2285.

101. Prelog, V.; Mutak, S. *Helv. Chim. Acta* **1983**, *66*, 2274.

102. Dobler, M.; Dumič, M.; Egli, M.; Prelog, V. *Angew. Chem.* **1985**, *97*, 793. *Angew. Chem. Int. Ed. Engl.* **1985**, *24*, 792.

103. Prelog, V.; Stojanac, Ž.; Kovačevič, K. *Helv. Chim. Acta* **1982**, *65*, 377. Prelog, V.; Mutak, S.; Kovačevič, K. *Helv. Chim. Acta* **1983**, *66*, 2279. Prelog, V.; Dumič, M. *Helv. Chim. Acta* **1986**, *69*, 5.

104. Prelog, V.; Jeger, O. *Biograph. Mem. Fellows R. Soc.* **1980**, *26*, 411. *Helv. Chim. Acta* **1983**, *66*, 1307.

105. Cornforth, J. W. In *Les Prix Nobel en 1975*; Nobel Foundation: Stockholm, 1976, p 34.

106. Ruzicka, L. *Bull. Soc. Chim. Fr.* **1928**, *43*, 1145.

107. Ruzicka, L. *Bull. Soc. Chim. Fr.* **1937**, *4*, (N.S.) 1301. See also *L'architecture des polyterpènes*. Special publication, October 1937.

108. Prelog, V. *Chemotherapia* **1963**, *7*, 133; *Experientia* **1964**, *20*(58), 336.

109. Prelog, V. *Pure Appl. Chem.* **1963**, *7*, 551.

110. Prelog, V. *Chimia* **1986**, *40*, 175.

111. Prelog, V. *Chimia* **1986**, *40*, 390.

112. Prelog, V. *Chimia* **1986**, *40*, 391.

113. Prelog, V. *Chimia* **1986**, *40*, 393.

114. The name rifomycin was coined by Sensi's collaborators when they saw the film *Rififi*. They did not know that another patented antibiotic was named rufomycin. Patent law demanded that names of two antibiotics had to differ in at least two letters; rifomycins were accordingly changed to rifamycins.

Index

Index

A

Academic posts
  at ETH, 31
  Falk-Plaut Lecturer (Columbia University), 48
  full professorship (ETH), 47
  laboratory directorship, 68
  postdoctoral studentship, 81
  Privatdocent, 31
  Reilly Lecturer (University of Notre Dame), 46
  University of Zagreb, 13
Acenaphthene, formation, 61, 62
Acetolysis, effect of ring size on rate, 59–60
Acklin, Werner, 65
Acyloin synthesis, use in ring cyclization, 39, 40*f*
Adamantane
  appearance on first day cover, 20
  attempted conversion to diamond, 19, 21
  discovery, 19–21
  synthesis, 19, 21
Adams, Roger
  anecdote, 91
  ETH visitor, 32
Alicyclic compounds
  key to reactivity differences, 38
  nonclassical strain, 39
Alkaloids
  biogenetic connections, 33
  joint projects at ETH, 33
  studies, 27–34
American Chemical Society, diamond jubilee, 47
Ancestors, 3
Androstanes, similarity to musk components, 38
Androstenes, 23–24
Androsterone, 23, 24
Anecdotes about fellow chemists, 90–93
Antibiotic–antagonist mixtures, 54
Antibiotics, 51–57
Antimalarial compounds, 12
Apprenticeship with Rudolf Lukeš, 9–10
Aricine, 33
Arigoni, Duilio, 68, 69 (photo)
Assassination of Archduke Franz Ferdinand, eyewitness, 3
Asymmetric cyanohydrin synthesis, 64
Asymmetric induction, 62
Asymmetric reactions, 62–64
Asymmetric synthesis, 48, 62, 64
Atrolactic acid synthesis, 64, 65
Austria–Hungary
  defeat in war, 6
  political situation in 1906, 4
Axial, proposed use of term, 78
Azulene, simple synthesis, 61, 62

B

Baeyer strain, 60
Bankankosine, 33
Barton, Derek H. R.
  alkaloid studies, 36
  as described by Prelog, 37
  conformational analysis, 41
  photo, 36
Batyl alcohol, discovery in mammals, 24
Beilstein's Handbuch, 13, 75, 77
Benzodiazepins, 22
Bergström, Sune, 23

Bicyclic amines, 12
Bicyclo[5.3.0]-1-azadecane, rearrangement, 17
Bielig, Hans-Joachim, 22
Biflorin, antibiotic from a plant, 56–57, 58
Biographical Memoirs of the Royal Society, 28, 85
Biologically active materials from animal organs, 22–24
Birth, 3
Boekelheide, Virgil, heteroalicyclic alkaloids, 33
Boromycin, 55, 56
Bosnia–Herzegovina, birthplace, 3
Bovet, Daniel, 15
Bredt's rule, 41–42
Bridgehead double bond
  formation, 41–42
  stability, 42
Bridgeman, Peter W., conversion of adamantane to diamond, 19, 21
Bromural (α-bromoisovaleryl urea), 90
Brown, H. C., 59
Brücke, Franz, 14
Brufani, Mario, 94
Buchanan, C., 75
Büchi, Georg, study of muscopyridine, 25–27
Burckhardt, Carl J., U.N. Commissioner, 71–72

C

Cahn, Robert S.
  assignment of descriptors to asymmetric centers, 75
  collaborator on CIP system, 74
  photo, 76
Campbell, Kenneth, 46
Carbonyl groups, stereoselective reduction, 64

Carboxyl group, general reduction method, 34
Carboxylic acid esters, rearrangements, 18
Caronia, 47
Centenary Lecture of the Chemical Society, 41
Cettolo, Mara (mother), 4 (photo)
Cevine, 34, 37–38
Charles University, 7
Chemical Engineering School of the Institute of Technology, 7
Chemical experiments
  at home, 5
  designed for Zagreb photographer in 1918, 6
Chemiker Zeitung, 7
Chemotherapeutic agents, 15
Childhood, 3–7
Chimyl alcohol, discovery in mammals, 24
Chiral bases, use as catalysts, 64
Chiral conformations, descriptors, 77
Chiral crown ethers, 82, 83
Chirality elements, 77
Chiroptical investigations, 79
Cholesterol alternatives, 28
Ciba–Geigy (CIBA or CIBA, Ltd.), source of financial support, 25, 31, 70
Cinchona alkaloids
  biogenetic connections, 33
  link to *Strychnos* alkaloids, 30
  synthesis, 16–18
Cinchonamine, 30, 33
CIP (Cahn–Ingold–Prelog) system
  acceptance, 78
  conceptualization, 74–75
  criticism, 76, 77
  major papers, 76–78
  supporting research, 78–79
  use in *Beilstein's Handbuch*, 77
  utility, 78

Citation Classics, 78
Cities visited in 1950–1951, 46, 47, 48
Citizenship, 74
Civetone, 23, 24, 38
Collaborators, 34–38
Collegial leadership, ETH laboratory, 69
β-Collidine, preparation, 18
Colloid mill, 39
Columbia University, chemistry faculty, 48
Competition in research, 34, 37
Conant, James, ETH visitor, 32
Configurational relationships between starting materials and products, general rule, 63
Conformation, first use of term, 41
Conformational analysis, 41
Conroy, Harold, Columbia faculty, 48
Constellational (conformational) effect, 39
Consultantship, Kastel, Ltd., 14
Coon, Minor J., 66
Coops, J., thermochemical studies of cycloalkanes, 61
Cope, Arthur C., nonclassical reactions, 57, 59
Cornforth, John, 64, 65 (photo), 86
Cornforth, Rita, 65 (photo)
Corynantheine, joint project, 33
Corynanthine, 33
Cram, Donald J., asymmetric synthesis, 47
Croatia, 3
Curvularia falcata
  isolation of oxido-reductases, 66
  reduction of cyclic ketones, 65–67
Cyclanols, acetolysis, 59
Cyclic compounds with 9–11 C atoms, preparation, 38–39
Cyclic esters, reaction with Mannich bases, 41
Cyclic ketones, microbial reduction, 65, 67
Cycloalkanes
  dehydrocyclizations, 61
  thermochemical studies, 61
Cycloalkenes, dehydrocyclizations, 61
Cyclodecane framework, structural analysis, 60
Cyclodecenes
  oxidation, 57, 59
  transannular reactions, 57
Cyclododecenes, classical behavior, 59
Cyclononenes, transannular reactions, 57
Cyclooctenes, transannular reactions, 57
Cyclopeptides, echinomycin, 53, 55
Cyclostereoisomerism, 79
Cycloundecenes, transannular reactions, 57
Czech, a new language for Prelog, 8
Czech Institute of Technology, 19
Czechoslovakia, unemployment, 11

D

Dalmatia, 3
Dauben, William, 47
Decalindiones, microbial reduction, 65, 67
Decalones
  microbial reduction, 65
  reduction by pig liver homogenate, 66
Dehydrocyclizations, 61
Desferrioxamines, differential binding to metals, 54
Diamond, attempted preparation from adamantane, 19, 21
Dihydroquinine, 12, 16
Diploma examination, 11

Djerassi, Carl, 89 (photo)
Doctoral examination, 11
Doering, William von E., Columbia faculty, 48
Driża, Gothard J., employer, 11–12
Duisberg, Carl, 90
Dunitz, Jack, 51, 60, 68, 70 (photo)
Dutler, Hans, 66
Dynamic Stereochemistry, symposium, 75

E

Ebel, Friedrich, 41
Echinomycin, 53, 55
Eidgenossische Technische Hochschule (ETH)
  first visit, 16
  modification of hierarchic structure, 69–70
  researchers, 50, 51
Elderfield, Robert, Columbia faculty, 48
Eliel, Ernest L., 46, 74 (photo)
Emigration to Switzerland, 21–22
Employment, first job, 11–12
Enantiomers
  separation by liquid–liquid partitioning, 81, 82, 84
  spontaneous resolution, 93
Enantioselective electrodes, 81–82
Engel, Bruno, 50, 68, 69, 87
Enzymic reactions, stereoselectivity, 64–68
Enzymic reduction of decalones, 66
Equatorial, proposed use of term, 78
Erni, Hans
  ex libris for Prelog, 88
  ex libris for Ruzicka, 87
Erysodine, 33, 34
Erythraline, 33, 34
Erythrina abessinica, 33

Erythrina alkaloids, 33
Eschenmoser, Albert, 68, 70 (photo)
Esters, configurational relations in reaction with Grignard reagent, 62–64
Estrogenic hormones, isolation from urine, 25
ETH, *See* Eidgenossische Technische Hochschule
Ex libris for Prelog, 88
Ex libris for Ruzicka, 87

F

Faraday Lecture of Wilhelm Ostwald, 8
Fatty acid synthetase complex, 66
Ferdinand Archduke Franz, murder, 3
Ferrimycins, 53–54, 56
Ferrioxamines, 53–54, 56, 57
Fieser, Louis, ETH visitor, 32
Financial success, 15
Fischer, Emil
  brilliant sugar chemist, 11, 75
  business sense (anecdote), 90
Fluoranthene, formation, 61, 62
Folkers, Karl
  Erythrina alkaloids, 33
  Merck researcher, 49
Force field method, 60
Fourneau, Ernst, 14
Freudenberg, Karl, 22
Full professorship at ETH, 47
Fürst, Andor, ETH researcher, 50
Furter, Max, director of Hoffmann–La Roche, 47

G

Gäumann, Ernst, phytopathologist, 51

Gelsemine, 33
Geometric enantiomerism, 79, 90
Gerlach, Hans, 79
German Chemical Society, 21
German occupation, 10
Glidden Company, patent challenge, 24
Glutamic acid, spontaneous resolution, 93
Glutamic acid synthesis, an anecdote, 93
Goldberg, Wolf Moses, 22, 23
Gonçalves, de Lima O., 55
Goutarel, Robert, ETH visitor and collaborator, 33
Grandparents, 3
Grignard reagent, reaction with esters, 62–64
Guest professorships, 73
Gunsalus, J. C., 64
Günthard, Hans, ETH researcher, 50

H

Habilitation, 50
Hammett, Louis, Columbia chemistry department head, 48
Hansley, V. L., 10-ring cyclization with colloid mill, 38
Hardegger, Emil, 50, 68
Harries, Carl ("Ozone"), 93
Harris, Stanton A., Merck researcher, 49
Haworth, Walter N., effects of ring size on sugar properties, 41
Heilbron, Ian, 37
Heilbronner, Edgar
  ETH colleague, 38, 50
  move to University of Basel, 69
  promotion, 68
  traveling on *Queen Mary*, 46
Heilmeyer, L., hematologist, 54

Helmchen, Günter, 78, 79
Heterocyclic compounds, 14–18
Heusser, Hans, ETH researcher, 50
Heyrovsky, Jaroslav, 42
High school
  Osijek, 7
  Zagreb, 5, 7
Hofmann, Fritz, 90
Home furnishings, damage by chemical experiments, 5
Hydride shifts, nonclassical, 60
Hydroquinones, effect of ring size on strain, 42
3-Hydroxy-20-oxo-5-pregnene, 24

I

Ibogaine, 33
Ibolutein, 33
Ile de France, 49
Illness, 13
Independent State of Croatia, 21
India, visit in 1963, 73
Industrial involvement
  Ciba–Geigy, 70, 71–72
  Kastel, Ltd., 14
  Merck, 49
Inferiority of chemist to X-ray crystallographers, 95
Influential teachers, 8, 9
Influential writers, 8
Ingold, Christopher
  assignment of descriptors to asymmetric centers, 75
  photo, 77
Institute for Special Botany (ETH), 51, 53
Institute of Fuels, 19
Instrumental revolution, consequences, 50–51
International Union of Chemical Societies, 16
Ion-specific electrodes, 81

Ionone, derivatives and rearrangement products, 25–26
Isolation procedures, refinement, 50
Isoprene rule, 51
IUPAC (International Union of Pure and Applied Chemistry)
 membership, 74
 Congress, 47

J

Jackman, Lloyd M., NMR studies, 57
Janot, Maurice Marie, ETH visitor and collaborator, 32
Jeger, Oskar
 coauthor of Ruzicka biography, 27–28
 photo, 70
 research collaborator, 34, 50, 68, 85
Jones, E. R. H. (Tim), 47

K

Kaarsemaker, S., thermochemical studies of cycloalkanes, 61
Kaiser Wilhelm Institute, 22
Kastel, Ltd. (now Pliva), industrial employer, 14
Keller, Walter, collaborator on antibiotic research, 51
Kelvin (Lord), 77
Kharash, Morris, ETH visitor, 32
Klyne, William, 78
Kohlbach, Dragutin, doctoral student, 15
Kuhn, Richard, 21, 22, 92
Kuria, Ivan, high school chemistry teacher, 7

L

Laboratory directorship at ETH, 68–70
Ladany, Eugen, 14
Landa, Stanislav, discovery of adamantane, 19
Lapworth, Arthur, 64
Large-ring compounds vs. medium-ring compounds, 41
Lecture tours, 73
Lectureships
 at Columbia, 48
 at Notre Dame, 45–46
Lepetit, Ltd. (also Lepetit S.A.), collaborator on rifamycins, 55, 91
Leprosy, effective drugs, 55
Leuchs, Hermann, 29
Librium, 22
London Chemical Society, 75
Lukeš, Rudolf, 9–10, 93
Lyle, G. G., 79
Lyle, R. E., 79

M

Mach, Ernst, influential writer, 8
Macrolides
 narbomycin, 52
 Prelog–Djerassi lactone, 52
Macrotetrolides, nonactin, 53
Magic with Koji Nakanishi, 73 (photo)
Major, Randolph, Merck research director, 49
Malaria, 13
Mannich bases, reaction with cyclic esters, 41
Manske, R. H., monograph, 34
Many-membered ring compounds
 dehydrocyclizations, 61
 microbial metabolites, 52

Many-membered ring compounds—*Continued*
  narbomycin, 52
  ring size vs. strain, 41
  thermochemical ring strain, 61
Marek, Ivan, professor of organic chemistry, 13
Marker, Russell, urine steroids, 25
Mass demonstrations, 4
McKenzie, Alexander, asymmetric synthesis, 62
McKenzie reaction, 63
Medicinal chemistry, 14–18
Medium-ring effect, 39, 40*f*
Medium-sized ring compounds
  abnormal behavior, 60–61
  nonclassical reactions, 57
  rapid acetolysis, 59–60
  studies, 38–45, 57, 59–62
Merck lectures, 49
Metacyclophanes, 42, 43, 44*f*
N-Methyl-2,5-diphenylpyrrole, triboluminescence, 11
Methylpyridines, rearrangements, 18
Microanalysis, difficulties, 10
Microbial metabolites, 51–57
Microbial reactions, stereoselectivity, 64–68
Microbial reduction of decalindiones, 65, 67
Military service, Royal Yugoslav Navy, 12–13
Moslems, 3
Mucor javanicus, oxido-reductases, 68
Muscone, 38
Muscopyridine, 25–27
Musk components, 38
Musklike compounds, 41

N

Nakanishi, Koji, 73 (photo)

Narbomycin, 52
Natural products studies, collaboration with Janot, 32–33
Negotiating skills, 78
New York, as described by Prelog, 48
Nitti, Filomena, 15
Nobel prize in chemistry, 86
Nomenclature problems, early awareness, 75
Nonactin, 53
Nonclassical hydride shifts, 60
Nonclassical reactions of medium-sized ring compounds, 57
Nonclassical strain, 60
Norlupinanes, 17
Notre Dame visit, 46

O

Oligomeric crown ethers, 82, 83
Oppolzer, Wolfgang, doctoral student, 55, 94
Organ extracts, studies, 22–27
Organic chemistry, changes in methodology, 49–51
Orthodox Jews, 3
Osijek, high school, 7
Ostwald, Wilhelm, influential writer, 8
Ottoman rule, 3
Ovchinnikov, Yuri, 79, 81 (photo)
Oxido-reductases
  in *Curvularia falcata,* 66
  in *Mucor javanicus,* 68
  stereoselectivity, 68

P

Palladized carbon, 61
Pallmann, Hans, ETH president, 47

Parents, 4 (photo), 5
Pasteur Institute, 15
Patent challenge, 24
"Pathological iron", 54
Pauling, Linus, 46
Phenolcarboxylic acids, synthesis, 34
Pig liver homogenate, reduction of decalones, 66
Pig pheromone, 23
Pitzer, Kenneth S., free rotation around C–C bonds, 38
Pitzer strain, 41, 60
Plattner, Placidus Andreas, 22, 38, 47, 48 (photo)
Pliva (formerly Kastel, Ltd.), 14
Poincaré, Henri, 8
Pope, William J., 93
Postdoctoral studentship, 81
Potato sprouts, source of cholesterol alternative, 28
Prague period
  chemistry practice, 12–13
  chemistry studies, 7–12
Praktikum, 8
Pregl, Fritz, 10
Prelog–Djerassi lactone, 52
Prelog, Jan (son), birth, 46
Prelog, Kamila (wife), 13, 49 (photo), 71 (photo)
Prelog, Milan (father), professional career, 7
Prelog, Vladimir (photos)
  as a sublieutenant in the Royal Yugoslav Navy, 13
  at 80th birthday, 70
  at Bürgenstock, 86
  discussing stereospecificity, 69
  in Sarajevo in 1912, 5
  with Charles C. Price, 45
  with Ernest Eliel, 74
  with Kamila Prelog, 49, 71
  with Karl Wiesner, 44
  with Koji Nakanishi, 73
  with Leopold Ruzicka, 27, 85

Prelog, Vladimir
  (photos)—*Continued*
  with Nobel laureates, 72
  with parents, 4
  with Woodward and Djerassi, 89
Preparative procedures, improvement in efficiency and selectivity, 50
Price, Charles C., 45 (photo), 46
Prieto, Augustin, 64
Professorship at ETH, 31
Prontosil (rubrum), chemotherapeutic agent, 15
Pseudoasymmetry, 79, 80
Pseudoyohimbine, 33
Publications
  biography of Ruzicka, 85
  chapters in *The Alkaloids*, 34
  Citation Classics, 78
  first paper, 7
  joint paper with Cornforth, 64
  papers on CIP system, 76–78
  review of alkaloids, 33
  "Specification of Molecular Chirality", 77

Q

Queen Elizabeth, 46
Queen Mary, 46
Quinine, partial synthesis, 12, 16
Quinones, effect of ring size on strain, 42
Quinotoxine, preparation, 16
Quinovic acid, 16
Quinuclidines, synthesis, 17

R

Rabe, Paul, dihydroquinine synthesis, 12, 16, 18
Ramirez, Fausto, Columbia faculty, 48

Rassi-maid (racemate), in design by Hans Erni, 88
Realgymnasium, 5
Rearrangements
  preparation of carboxylic acid esters, 18
  preparation of methylpyridines, 18
  synthesis of bicyclo[5.3.0]-1-azadecane, 17
Reduction
  by *Curvularia falcata*, 65–67
  by pig liver homogenate, 66
Reichstein, Tadeusz, hormone isolation, 23
Research
  adamantane, 19–21
  alkaloids, 27–34
  androstenes, 23
  antibiotics, 51–57
  asymmetric reactions, 62–64
  at ETH, 22–34
  cevine, 34, 37–38
  change in direction, 52
  cholesterol alternatives, 28
  Cinchona alkaloids, 12, 16–18
  collaborators, 34–38
  crown ethers, 82
  during World War II, 16
  first research topic, 10–11
  graduation project, 11
  heterocyclic compounds, 14–18
  ionones, 25–26
  medicinal chemistry, 14–18
  medium-sized rings, 38–45, 57, 59–62
  microbial metabolites, 51–57
  norlupinanes, 17
  organ extracts, 22–27
  partial synthesis of quinine, 16
  poaching on Rabe's ground, 16
  quinovic acid, 16
  quinuclidines, 17
  rearrangements, 17–18
  resolution of Tröger base, 30
  separation of enantiomers, 81
  solanine, 28
Research—*Continued*
  stereochemical studies in connection with CIP system, 78–79
  stereoselectivity of asymmetric syntheses, 64–68
  strychnine, 29–31
  sulfanilamide–azo dyes, 15
  transannular reactions, 57, 59–62
  work with Ruzicka, 16
Retirement, 81
Rhamnoconvolvuline, determination of constitution, 11
Richter, Friedrich, 75, 77
Rifampicin, 55
Rifamycin, the story, 94–95
Rifamycins, 55, 58
Rifomycin, *See* Rifamycin
Ring size, effect on strain, 41–42
Robinson, Robert
  criticisms of CIP system, 76
  dislike of CIP system (anecdote), 91
  ETH visitor, 32
  studies of strychnine structure, 29–31
  with Lady Robinson on the day after their wedding, 29 (photo)
Rockefeller Foundation, 22
Rosenkranz, Georg, 22
Royal Yugoslav Navy, 12–13
Rubrum (prontosil), chemotherapeutic agent, 15
Russian chemists, limited contact, 79
Ruzicka, Leopold
  art collection, 31
  biography, 28
  Croatian origin, 16
  ex libris by Hans Erni, 87 (photo)
  friend and patron, 27
  in Paris (anecdote), 92
  obtaining Swiss visas for Prelogs, 21
  photos, 27, 85

Ruzicka, Leopold—*Continued*
  plan for Woodward Institute, 72
  postwar activities, 31
  relationship of hormone and pheromone structures, 23

S

Sarajevo
  memorable events, 3
  people, 3
Sarett, Louis, Merck researcher, 49
Schenk, Günther O., 22
Schleyer, Paul von Ragué, adamantane, 21
Schooling
  chemistry studies in Prague, 7–12
  Czech Institute of Technology, 7–12
  high school, 5–7
Seebach, Dieter, 78
Seibl, Josef, mass spectrometrist, 51
Selye, Hans, stress compounds, 24
Sempervirine, 33
Sensi, Piero, collaborator on rifamycins, 55, 94
Serbs, 3
Sex hormones, industrial synthesis, 24
Shemyakin, Mikhail, Russian natural products chemist, 79
Shoppee, Charles, 47
Simon, Wilhelm, molecular spectroscopist, 51, 68, 81, 84 (photo)
Slavic peoples, unification, 6
Slavonia, 3
Snatzke, Günther, 79
Solanidine, structure determination, 28
Solanine, studies, 28
Solanum alkaloids, 34
Soltys, George, solanidine structure, 28
Spanioles, 3
Specification of Molecular Chirality, 77
Sputnik, 69
Staudinger, Hermann, 92
Stereochemical problems, 78–84
Stereoisomer specification, *See* CIP system
Stereoisomers, need for unambiguous specification, 75
Stereoselective reduction of carbonyl groups, 64
Stereoselectivity of microbial and enzymic reactions, 64–68
Sternbach, Leo, 22
Stoll, Max, 39
Strain, nonclassical, 60
Streptazol, 15
Streptomyces metabolites, studies, 52
Structure determination, new methods, 50–51
Strychnine, structural studies, 30
Strychnos alkaloids, link to *Cinchona* alkaloids, 30
Sulfanilamide, chemotherapeutic agent, 15
Sulfanilamide–azo dyes, 15
Swiss citizenship, 74
Syntex, Ltd., 22
Szpilfogel, Stefan, collaborator, 30

T

Tartaric esters, 82, 84
The Alkaloids, 34
The School of Chemistry, treasured book, 8
Thermochemical ring strain, 61
Thermochemical studies of cycloalkanes, 61
Thiol esters, reduction, 34
Todd, Alexander, 47, 73
Tollens, Bernhardt, 9

Transannular reactions, 57, 59–62
Travels, 73
Triboluminescence, 11
Trihydroxamic acid–iron(III) complexes, 54, 56
Tröger base
  resolution to enantiomers, 30
  structure, 31
Truffles, chemical explanation for search by pigs, 23–24
Trust in research, 37
Tuberculosis, effective drugs, 55
Tulinsky, Alexander, 94

U

Unemployment, 11
University of Zagreb (1935–1941), 13–21
Upbringing by aunt, 5
Urine, source of estrogenic hormones, 25

V

Vaciago, Alessandro, 94–95
Valium, 22
van Kamp, H., thermochemical studies of cycloalkanes, 61
Veratrum alkaloids, 34
Verkade, Peter, IUPAC chairman, 74
Vespirenes, 79, 81
Vitek, Kamila (wife), *See* Prelog, Kamila
von Halban, Hans, 38
Votocek, Emil, professor of organic chemistry, 9, 93

W

Wald, Franz, influential lecturer, 8

Wallenfels, Kurt, solanidine structure, 28
War against Serbia, 3–4
Weissback, Otto, 77
Weizmann Institute, 91
Wessely, Fritz, 94
Westheimer, Frank, 69 (photo)
Wettstein, Albert, CIBA director, 72, 94
Wieland, Theo, 22
Wiesner, Karl, polarographer, 42, 44 (photo)
Wilhelm, Max, 64
Wilson Laboratories, 22
Wintersteiner, Otto, 23
Wöhler, F., hematologist, 54
Woodward Institute, 72
Woodward, Robert Burns
  alkaloid studies, 36
  as described by Prelog, 36–37
  in India, 73
  personal friend, 37
  photos, 35, 89
  reflection about Robinson (anecdote), 91–92
  strychnine studies, 30
  Swiss laboratory, 72
World War II
  departure of Jewish scientists from ETH, 22
  diversions caused by war, 29
  effect on research, 28–29, 32
  research climate, 16

X

X-ray crystallography, competition with chemical structure determination, 94–95

Y

Yohimbine, 33

Youth, 3–7
Yugoslavia
  unemployment, 11
  unification, 6

Z

Zagreb, high school, 5, 7
Zähner Hans, ferrioxamines, 53

Production: Peggy D. Smith
Copyediting and Indexing: A. Maureen Rouhi
Cover: Tina Mion
Acquisition: Robin Giroux

Books printed and bound by Maple Press, York, PA

Paper meets minimum requirements of American National Standard for Information Sciences—Permanence of Paper for Printed Library Materials, ANSI Z39.48–1984 ∞